Dual Water Systems

AWWA MANUAL M24

First Edition

American Water Works Association

Denver, Colorado

Printed in USA

ISBN 0-89867-290-2

Table of Contents

Foreword

This publication contains information on some existing and theoretical dual water systems. The American Water Works Association does not endorse any of the information or systems contained herein. This publication is not presented as a recommended standard of practice, but as a reference book for reporting the state of the art. The Distribution Division Committee on Dual Distribution Systems views dual water systems as a valuable management tool for communities that have limited supplies of high-quality water treatable to the federal drinking water standards. Despite the lack of nationally accepted criteria and guidelines, the committee recognizes that properly treated and distributed nonpotable water, as defined herein, can safely be used for irrigation, industrial, and other purposes outside the home. This practice will help preserve the precious limited high-quality water sources for drinking, cooking, and other uses inside the home. The committee believes that a national program of research is needed before a standard of practice on dual water systems can be published. This publication is offered as a starting point toward the development of such a standard of practice.

Acknowledgments

This manual was prepared by the Distribution Division Committee on Dual Distribution Systems. The membership of the committee at the time it approved this manual was as follows:

Lloyd A. Dove, *Chairman*
Arun K. Deb, *Vice Chairman*

Members

W.J. Boegly
J.T. Cookson Jr.
James Crook
Michael P. Evans
Richard Fenton
P.D. Haney
R.D. Lee
D.A. Okun
A.F. Vondrick
P.E. Pilgreen*

*Distribution division liaison

Introduction

Dual distribution systems for community water supplies may become common practice in the next century. Diminishing supplies of quality resources and rapidly escalating costs of treating both potable water and wastewater are the two main reasons to believe so.

Community water managers and water system design engineers should concentrate on planning, developing, and implementing dual water systems to serve all the needs of a typical community. In decades to come it may be cost prohibitive to treat a quality source for uses that do not require potable water.

Although dual systems is a popular contemporary idea, the concept and practice of more than one delivery system is not modern-day. The first known dual distribution system for water supply was that built by Augustus in Rome almost 2000 years ago. According to Frontinus,[1] a new aqueduct carried water that was so unwholesome it was not used for potable purposes. The water was used in order "to avoid drawing upon the better sources of supply. . . ." And in 1911, W.T. Sedgwick of the Massachusetts Institute of Technology read a paper to the New England Water Works Association entitled "Has the Time Come for Double Municipal Water Supplies, One—Naturally Pure—for Drinking and Cooking, the Other—Denatured or Sterilized—for All Other Purposes?"[2]

Preserving limited quantities of high-quality water for potable use should be the reason for developing dual distribution systems for water supply; one system for potable water, the other for nonpotable. A policy affirming this approach was enunciated by the United Nations Economic and Social Council: "No higher quality water, unless there is a surplus of it, should be used for a purpose that can tolerate a lower grade."[3]

With increasing urban and industrial development, and with most high-quality sources with substantial yield already developed, surpluses of high-quality water for the future are becoming less available and, in many areas, have already been depleted.

The utilization of wastewater treatment plant effluents for industrial or urban irrigation purposes was one approach to developing a nonpotable supply to relieve the demand on potable water. An example is the introduction of secondary effluent in Baltimore, Md., 50 years ago for use in the city's steel industry.[4] Another is the distribution of reclaimed wastewater to several industries in a community such as was adopted in Colorado Springs, Colo., in 1960 to relieve the demand on limited freshwater resources.

Two factors that should accelerate the development of dual distribution systems are: (1) requirements for great amounts of water that do not necessarily need to be of potable quality for industrial use and urban irrigation, particularly in arid areas, and (2) requirements for costly advanced wastewater treatment, including nutrient removal, to protect receiving waters.

If reclaimed wastewater is to be used for urban irrigation, nutrient removal becomes not only unnecessary but wasteful, since nutrients already in the wastewater can replace those that would ordinarily need to be purchased. Because urban irrigation, including the watering of parks, campuses, median strips, and residential lawns, offers opportunities to use reclaimed wastewater, a distribution system is required to permit the easy use of wastewater for other nonpotable purposes.

Although many other dual distribution systems of varying degrees of complexity have been built, neither state nor federal governments have yet established guidelines for nonpotable distribution systems. Thus, this manual represents the state of the art for dual distribution systems so that experiences with systems now in operation may be utilized by others contemplating such systems. The manual can serve as a predecessor to more formal regulations and standards that may be prepared at either the state or federal level.

This manual should be a useful beginning for addressing these issues. Because the dual distribution practice is barely developed, this manual can be considered to be dynamic. Hopefully, users of the manual will add their experience to the next edition.

References

1. FRONTINUS, S.J. *The Water Supply of the City of Rome* (trans. by Clemens Herschel). NEWWA (1973).
2. SEDGWICK, T.W. & LETTON, H.P. Has the Time Come for Double Municipal Water Supplies, One—Naturally Pure—for Drinking and Cooking, the Other—Denatured and Sterilized—For All Other Purposes? *Jour. NEWWA*, 25:400 (1911).
3. United Nations Economic and Social Council. Water for Industrial Use Rept. E3058 ST/ECA/50 (1958).
4. OKUN, D.A. Planning for Water Reuse. *Jour. AWWA*, 65:10:617 (Oct. 1973).

Chapter 1

Overview

This chapter will consider the areas in which dual distribution systems would be potentially advantageous. It will also touch on the current state of governmental regulations for dual distribution systems and will introduce terminology to be used throughout the rest of the book.

Potential Applications for a Dual Distribution System

Most existing dual distribution systems are the result of the initiative of one or two individuals who, seeing the potential advantages, seized the opportunity and were successful. With many successful systems now in operation, varying from systems that serve one or two major customers to systems that serve every property in a community, municipal officials and their consulting engineers should consider dual distribution systems as among the options available to them for addressing water supply needs, water pollution control problems, or both. Following are some situations in which a dual distribution system would appear to be worthy of evaluation.

Limited water supplies. The yield of the present water supply is expected to be exceeded by demand, and an additional supply needs to be developed. Consideration could be given to the gradual introduction of a nonpotable system. The advantage of such an approach would usually be economic, since new supplies usually cost more to develop than existing supplies. The original source of supply was probably selected as being lower in cost than other sources because of transmission distances or more extensive supply or treatment facilities. Water users, whether industrial, commercial, or municipal, may well be served with reclaimed wastewater in place of potable water at a cost substantially lower than the cost of developing new sources of supply for potable purposes.

Limited water resources. In many cases competition for existing water resources or limitations on these resources make acquisition of additional water resources politically and financially difficult. For example, obtaining additional water resources might require interbasin transfer, which is less politically acceptable now than it was in the earlier years of water resources development. Greater knowledge of groundwater hydrology has illuminated the danger of excessive withdrawals from aquifers, which can result in continually higher costs of pumping, impaired quality, and possibly land subsidence. Constrained resources create a situation where water reclamation and a dual distribution system become options.

Polluted sources. The 1962 Public Health Service Drinking Water Standards recommended that the "water supply should be obtained from the most desirable source which is feasible. . ." The 1976 US Environmental Protection Agency (USEPA) National Interim Primary Drinking Water Regulations (NIPDWR) state that ". . .priority should be given to the selection of the purest source." Many communities have selected the highest quality source available, and additional development may require use of a more polluted source. In the past, engineers properly believed that they could render almost any polluted water safe for potable supply through adequate filtration and disinfection. However, "chemical revolution" following World War II has resulted in the creation of thousands of stable synthetic organic chemicals that are not readily degraded in the environment or easily removed by treatment. The treatment of water that will meet present and future drinking water standards will cost more. Contaminated supplies may be excluded as a source of potable water but safely used for nonpotable purposes where chemical contamination is of little significance.

Rigorous wastewater treatment. Because of the goal of restoring and preserving the quality of the nation's waters, treatment requirements for community wastewater have become exceedingly stringent and costly. In many cases nutrients may need to be removed. They have little health significance but are troublesome because they enrich receiving waters and cause eutrophication. The resulting algal blooms make water more difficult to treat. If the wastewater can be recycled for industry or for urban irrigation, then the advanced waste treatment, including nutrient removal, may not be necessary and may in fact be undesirable. The potential savings in costs of wastewater treatment have not, up to now, been sufficient to encourage reuse because federal construction grants pay 75 percent of the total capital cost and state subsidies cover another 10 to 15 percent. However, many communities have begun to appreciate that the operation and maintenance of advanced wastewater treatment (AWT) facilities is exceedingly costly and that, if the wastewater can be used productively, the savings in wastewater treatment might be passed on to users, making water reclamation and dual distribution systems that much more economically attractive.

The following are some applications and advantages of dual distribution systems:

1. Raw water supply mains between the supply source and the treatment facility can be tapped and the water utilized with or without added treatment.

2. Separate distribution systems from various water supply sources can be constructed for power plant cooling, urban irrigation or lawn sprinkling, incinerator air-pollution control and washdown, or industrial processes.

3. Treated wastewater can be distributed in separate pipelines within the

treatment plant site to satisfy the demand for chlorinator water, site irrigation, odor control devices, plant washdown, and firefighting. Treated wastewater can also be taken from plant outfalls and reused, with or without additional treatment.

4. Treated wastewater can be distributed through separate pipelines. The degree of wastewater treatment will depend on the intended end use; e.g., lawn sprinkling with wastewater where potential public exposure to the spray or aerosols exists should require virus inactivation in the treatment process.

5. Nonpotable water can be supplied in separate pipelines under high pressure to meet fire demands. The source of this nonpotable water can be groundwater, surface water, or treated wastewater.

6. Abandoned pipelines may be utilized to transport nonpotable water to areas where it is needed. Obviously, thorough examination, testing, and cleaning of such lines must be performed to assure their suitability for the intended purpose.

7. Treated wastewater, groundwater, or surface water that is unsuitable for treatment to potable standards may be transported by separate pipelines to agricultural or industrial areas where lower quality water is acceptable.

8. Existing parallel water supply lines offer the potential for upgrading water quality in one line to meet the USEPA NIPDWR requirements and using the other line for subpotable or nonpotable requirements.

Guidelines

Standards or regulations for dual water systems have not yet been promulgated by the federal or state governments. Criteria for water reclamation have been established by some states. The most comprehensive standards are the Wastewater Reclamation Criteria of the state of California.[1] These criteria are directed toward treatment trains and standards in the treatment process for a wide variety of uses, including contact recreation, which calls for the highest standard. Heavy emphasis is placed on the reliability of treatment. Presumably, if contact recreation is one of the uses of water from a nonpotable distribution system, then that water should be suitable for other nonpotable uses as well.

The standards that have been used for nonpotable systems have been adopted by the agencies responsible for the systems themselves, such as the city of St. Petersburg (Fla.) and the Irvine (Calif.) Ranch Water District. The latter has published the most comprehensive standards and regulations.[2, 3]

The USEPA has published Guidelines for Water Reuse,[4] which offers useful information to those interested in the potential for water reuse, including information applicable to dual distribution systems.

To date, however, the USEPA, AWWA, and state water quality standards, regulations, specifications, and criteria developed for potable water supply systems have not been formulated for nonpotable systems. In fact, because the distribution of reclaimed water is generally the responsibility of the sewerage authority in a community, the local approach has not been as responsive to health concerns as would be the case if the reclaimed water were viewed as part of the community water supply system.

Another complication is that viruses are the principal health concern arising from the distribution of a nonpotable water, especially one derived from wastewater. Where nonpotable waters are used for a single application, exposure is limited to a relatively small number of people. When those waters are distributed widely through a community, the exposure may be quite high, such as where wastewater is sprayed for urban irrigation or used for cooling towers. Virus standards have not yet been adopted for potable waters. Accordingly, the approach to viral safety continues to be dependent on bacterial measures and chlorination.

Among the standards, guidelines, criteria, and regulations still needed for dual distributions systems are the following:

- monitoring and standards requirements for turbidity, bacteria, and chlorine residual at the source (the treatment plant effluent for reclaimed wastewater) and at points through the distribution system;
- reliability requirements, including standby provisions and automatic wasting of water not meeting standards;
- personnel requirements;
- cross-connection prevention requirements;
- plumbing codes for use of nonpotable water;
- materials requirements for piping, valves, hydrants, and other fixtures to distinguish nonpotable from potable;
- storage facility requirements to permit holding excess nonpotable waters during wet seasons; and
- procedures for use of nonpotable waters, such as restricted hours of application.

Definitions

In the context of this publication, the following terms and meanings apply.

Dual water distribution system (dual system): A water distribution facility that distributes two grades of water to the same service area; one potable and the other perhaps nonpotable. The quality, quantity, and pressure available in each system are functions of the sources and intended uses for each grade of water. The dual systems may be limited or complete, as defined hereafter.

Limited dual water distribution system: Consists of one potable water system that covers a total service area and one nonpotable water system that covers a limited area.

Complete dual water distribution system: Two systems, one potable and one nonpotable, covering a total service area. The nonpotable water system is utilized for all applications that do not require potable water, without risk of cross connection or inadvertent use, based on health and water quality considerations.

Potable water: Water of excellent quality intended for drinking, cooking, and cleansing uses. This grade of water would conform to the water quality requirements of state and federal regulatory agencies.

Nonpotable water: Water acceptable for other uses than potable. This water would be safe for occasional inadvertent human consumption and would generally conform to the water quality requirements of state and federal regulatory agencies for nonpotable water.

References

1. California Admin. Code. Water Reclamation Criteria. Title 22, Div. 4 (1978).
2. Irvine Ranch Water District. Rules and Regulations for Water, Sewer, and Reclaimed Water Service. Calif. (1977).
3. Irvine Ranch Water District. Standard Specifications for the Construction of Water, Sewer, and Reclaimed Water Facilities. Calif. (1979).
4. Camp, Dresser & McKee, Inc. USEPA Guidelines for Water Reuse. (August 1980).

Chapter 2

Experiences With Dual Distribution Systems

As indicated in Chapter 1, the utilization of water supplies that have different levels of quality for specific end uses is not a new idea. Many cases are reported in the literature in which wastewater is reclaimed for special purposes—such as urban irrigation, power plant cooling, and industrial consumption—that require some form of dual distribution system to maintain separate potable and nonpotable water supplies. Unfortunately, many of the articles are sketchy and tend to emphasize only the use of the reused water. No article describes a complete dual distribution system. This chapter will provide some indication of where and to what extent dual distribution systems have been installed.

United States

No literature was found that describes extensive dual distribution systems in the Northeast region of the United States, although a number of articles indicate significant problems in locating adequate supplies of water. The application of dual distribution systems has been proposed for a number of shopping malls in the Washington–Baltimore area.[1] Each mall is a completely enclosed facility consisting of 700 000 to 1 million sq ft (65 000 to 93 000 m^2) of commercial space. A water recycling and distribution system has been proposed for each of these malls based on the particular needs of the developer and the economic feasibility of integrating the various utility services in a mall complex.

Under certain conditions the total utility system can be implemented in a manner to provide complete reuse of the reclaimed effluent, eliminating discharge to a receiving stream or sewerage system. Several similar systems have been proposed by private developers in the state of Maryland. An impetus for the application of recycle systems is the moratorium on new sewer connections imposed in portions of the state. The water reclaimed from treated effluent would be recycled for flushing-toilet facilities; electrical generation, heating, and cooling needs; and scrubbers in the cleaning of flue gases.

If the water and energy demands for a commercial complex balance, it is possible to completely reuse the effluent generated in the facility. The US Department of Housing and Urban Development has conducted extensive research on the application of Modular Integrated Utility Systems (MIUS) in which reuse and recycling were applied to housing developments and commercial facilities.[2,3]

The largest single project proposing a dual distribution system in the eastern United States is the International Trade and Convention Center, which provides multiple services to the nation's capital. It will include a trade center, exposition center, cultural center, convention center, sports facility, Olympic center, recreational center, and medical research center.

The total scope of this facility can be recognized when one considers it will occupy 1000 acres (400 ha) and include a 1150-ft (350-m) diameter domed stadium with adjoining golf courses, fairgrounds, Olympic grandstands, 2500 hotel and apartment units, and an international medical center. The International Trade and Convention Center will have on-site utility facilities that will include an electric power generating station, wastewater treatment with water recovery and reuse, and solid-waste handling facilities.

Storm and groundwater will be purified for either potable or nonpotable supplies depending on demands, and liquid waste will be purified and used as nonpotable water. The nonpotable water will be used primarily for maintenance of buildings and grounds, and integrated with the electrical power system to provide water for heating and cooling and steam generation. County or municipal water supplies will serve as a back-up to replenish the system.

In January 1976 a one-page questionnaire was distributed to 326 cities with populations of 10 000 or more in the 10 southeastern states of Alabama, Arkansas, Florida, Georgia, Kentucky, Louisiana, Mississippi, North Carolina, South Carolina, and Tennessee.[4] Of the 191 responses (58.6 percent), only 10 of the cities had existing segments of a dual distribution system, and only seven additional cities were planning such systems.[5]

St. Petersburg, Fla., has one of the largest dual distribution systems in the United States. Details on the design and construction of the St. Petersburg system, along with information on the method of uniquely marking fire hydrants, valve boxes, and pipelines are presented in Chapters 4 and 5.

The survey also indicated that the city of Tuscaloosa, Ala., draws untreated water from a raw water pipeline for use in supplying industries located along the route of the raw water intake line.

Texarkana, Ark., also has a nonpotable water system that supplies water to a local industrial plant. This system consists of three shallow wells and 1.5 mi (2.4 km) of 6-in. (150-mm) water main. The average water consumption for this system is 225 000 gpd (0.85 ML/d).

Several applications of dual distribution systems in the arid areas of the southwest have been described by Vondrick.[6] Because of the increasing demand for domestic, industrial, and agricultural water use and the limited water supplies available, the states of Colorado, Arizona, and California have reported considerable interest in the application of dual distribution systems.

Grand Canyon National Park installed a dual system in the 1920s because the freshwater source was limited and located at the bottom of the canyon. Cabins in the park have two sets of plumbing. Reclaimed wastewater is used for the nonpotable supply of water and the lines are painted red.

The dual distribution system in the Phoenix, Ariz., area is unusual in that the nonpotable system is composed of canals and the water is used for irrigation purposes. As a result there is no particular problem associated with separating the two systems since the nonpotable system is gravity fed and the municipal water system is under pressure. Most of the interference between the two systems occurred during construction, but cross connections do not appear to be a problem of concern to the potable water system operators. In some areas the canals parallel the streets in residential areas, allowing the residents to obtain irrigation water without difficulty. Phoenix has also made arrangements to sell effluent from its treatment plant to an irrigation company.

The Irvine Ranch Water District in Orange County, Calif., has constructed an urban dual distribution system that is similar in many ways to the system described for St. Petersburg.[7] In the Irvine district the uses of reclaimed water include landscape and agricultural irrigation, water for construction purposes, industrial process water, and recreational impoundment. As of 1978 the system was supplied by a 15-mgd (57-ML/d) activated sludge treatment plant with filtration and disinfection followed by storage in two 2000 acre-ft (2.5×10^6 m^3) reservoirs. The distribution system was installed in three villages in 1978 and was planned for all new development.

The Colorado Springs, Colo., system consists of two pressure lines running east and west of the tertiary wastewater reclamation plant. Both lines terminate at abandoned water reservoirs, from which several sites are irrigated. Blue and white fire hydrants are located on the pressure lines.[8] Irrigation sites include 27 acres (11 ha) for the wastewater reclamation plant, 125 acres (50 ha) of parks, 206 acres (83 ha) of cemeteries, 317 acres (128 ha) of golf courses, and 105 acres (43 ha) of institutional land. Tertiary treatment was added to the system in 1971 and 336 mil gal (1271 ML) were treated by this process; 203 mil gal (768 ML) were used for irrigation and 133 mil gal (503 ML) for industrial water. The charge to users was 30 cents per 1000 gal (7.9 cents per 1000 L) plus pumping.

Denver, Colo., is also proposing to reclaim municipal waste effluents for industrial and agricultural use. A 1-mgd (3.8-ML/d) advanced waste treatment plant is designed to produce water of drinkable quality. This effluent will be directed to a recreational lake until data have been collected to determine if the treated effluent should be mixed with the existing municipal water supply.[9] There is some indication in the literature that consideration is being directed toward the possible use of this effluent for other than drinking purposes. If so, a dual distribution system similar to that in St. Petersburg would be required.

International Experience

International experience on the use of dual distribution systems is not well documented. In an attempt to obtain information on foreign systems, a limited survey was made by contacting 11 organizations that were described as research organizations on water supplies in the *European Research Index*.[10] Only six replies were received. No reports were found that described the installation of dual distribution systems in foreign countries, but a number of articles indicated that dual systems might be in use. In general, any articles dealing with reuse or reclaiming of treated waste effluents were assumed to require some form of dual piping to provide separate distribution of potable and nonpotable water.

Many European countries have expressed a concern with the increasing degree of water pollution resulting from industrial waste discharges.[11,12] Many of the articles concern only small local problem areas, but some of them address the water needs of a country as a whole. As government policies on the reduction of pollutants in municipal and industrial discharges require that a greater degree of waste treatment be provided, considering the associated increased costs, it appears that these highly treated effluents represent a valuable water resource. Such effluents could be utilized rather than discharged to the nearest body of water itself, which may contain higher concentrations of pollutants than the treated wastes. Other than in magnitude, international and US problems are much the same.

Italy. As far as can be determined, it appears that one of the first dual distribution systems (as defined by the AWWA Dual Distribution Committee) was installed in Rome in the period 40 to 103 AD.[13] Some of the aqueducts were reported to convey water considered of inferior grade to baths, washing establishments, and for watering gardens.[14] The remaining aqueducts were used for drinking water supplies. Certain of the fountains of Rome were also, and still may be, supplied from a nonpotable source. It is possible that similar systems were also provided in other European countries during this period but are not recorded in literature.

In a reply to the survey mentioned earlier, R. Passino described the dual distribution system that has existed in Rome for the past 30 to 40 years.[15] Passino indicated that the dual systems were installed because nonpotable water systems required lower pressures and could be used in areas where drinking requirements were limited (parks, green zones, and service areas). The city was divided into zones based on areas to be served and elevation, with the primary supply lines ranging in size from 12 to 14 in. (300 to 350 mm) in diameter.

In the 1960s, a new plan was advanced to expand the nonpotable system to a proposed capacity of 250 mgd (945 ML/d). Of this amount, about 160 mgd (605 ML/d) would be for industrial supplies, 57 mgd (215 ML/d) for lawn sprinkling and street washing, and 33 mgd (125 ML/d) for other uses. The first section of this dual network is currently under construction and has a potential capacity of 11.4–18.3 mgd (43–69 ML/d).

Passino also included a brief description of a dual network serving the Vatican City. Nonpotable water is drawn from the Paolo Aqueduct at Piazza Carpegna at a rate of about 1270–1590 gpm (80–100 L/s) and is used for water closet cisterns, laundries, thermoelectric power plants, and air conditioning, thus reducing potable water consumption to a minimum.

West Germany. The reuse of wastewater from one industry to another was suggested in 1969; however, it was pointed out at the time the article was prepared that there were no installations of this type in West Germany.[12] In a later report

Shenderuk indicated that current emphasis was being directed to reducing or eliminating water withdrawals by the installation of closed-circuit recycling systems within industry, or to reducing water pollution by using treated effluents for irrigation.[16] Shenderuk also anticipated that greater use of treated wastewater will be made in the future, which would require the installation of dual distribution systems to keep potable water separate from nonpotable water.[16]

South Africa. Stander and Funke have described the need for water reuse in South Africa.[17] The Windhoek sewage purification works produces a high quality effluent that is blended with water from existing supplies and is used as the source of potable water. A dual distribution system is not used. The authors also describe installations in which treated effluents are used for cooling water for power plants and for various types of industrial process waters. It can be assumed that dual distribution systems are employed for these applications, although they are not described in the article.

Laburn discusses various effluent reuse possibilities and also mentions the use of dual distribution systems for water management planning in the Pretoria metropolitan area.[18] He points out that the use of such systems had been ruled out in the past because of the costs involved; however, as the costs of raw water and purification increase, the potential use of such systems becomes a more real possibility.

France. There is a dual distribution system in Paris, with portions of the system installed in the Paris sewers.[19,20] An article in *Engineering News*[21] in 1910 described the walk-through combined sewer under the Boulevard Sebastopol, which contained both a 42-in. (1.07-m) drinking water supply main and a 36-in. (0.91-m) river water supply pipe which was used to supply water to manufacturers and for other industrial purposes. An interesting feature of this combined sewer design was the inclusion of a smaller diameter pipe for spring water. Unfortunately, little information can be found in the literature on the current status of the Paris dual distribution system.

United Kingdom. Although it is not listed as a "general practice" in England and Wales, in some cases the local water company does operate a separately piped supply of industrial-grade water.[22] Deb and Ives[23] have reported that there are about 42 dual water supplies in England and Wales.

Union of Soviet Socialist Republics. Ryabyshev[24] has discussed the water supply for Moscow and has pointed out that the large amount of potable water used by industrial plants could be replaced with river water having no treatment or minimal treatment. Ryabyshev also describes a network of industrial water supplies that use river water to supply industrial plants (mainly for cooling) in which the main treatment is the removal of coarse particles. Two of these systems are in existence and have a combined capacity of about 205 mgd (776 ML/d). A third system that will utilize treated wastewater is under construction. Anticipated capacity of this system is 115 mgd (435 ML/d), bringing the total capacity of the three systems to 320 mgd (1211 ML/d).

Mexico. Griffith has described a number of industrial installations in which treated effluents are reused for process water, cooling water, and irrigation.[25] A list of plant capacities and the ultimate use of the effluent is given for 15 installations in Mexico. One of the reasons cited for water reuse is the lack of available water resources. Dual distribution systems are not mentioned.

Other countries. Schmidt, Beardsley, and Clements[26] have tabulated 15 installations where reuse is practiced overseas. All of the systems appear to be treated wastewater. Okun[27,28] has mentioned dual systems in the Bahamas, Bangkok, Singapore, and parts of Hong Kong.

Summary

The purpose of this chapter was to determine the extent to which dual distribution systems are used in the United States and foreign countries. By using surveys and literature reviews it was possible to locate a few cities that have some form of dual distribution system currently installed or planned. The reasons most commonly cited for not installing such systems to a greater extent were concerns with the potential health aspects of using reclaimed water for potable purposes and the costs of installing such systems, especially in areas where municipal water systems are already installed. Although the data are sketchy there appears to be an increasing interest in reusing treated wastewaters as a water supply to complement the potable water currently being distributed in cities.

References

1. Cookson, J.T. Jr. Dual Distribution Systems in the Northeastern United States. Proc. AWWA Dual Distribution Systems Seminar. Denver, Colo. (June 26, 1976).
2. Samuels, G. & Boegly, W.J. Jr. MIUS Technology Evaluation: Preliminary Subsystem Recommendations for Near-Term Concepts. ORNL/HUD/MIUS-17 (1976).
3. Boyle, J.W. et al. Plan for Environmental Evaluation of MIUS Demonstration at St. Charles, Maryland. ORNL/HUD/MIUS-51 (1978).
4. Dove, L.A. Dual Distribution Systems in the Southeastern United States. Proc. AWWA. Dual Distribution Systems Seminar. Denver, Colo. (June 26, 1976).
5. Dove, L.A. Application of Dual Distribution Systems. Proc. AWWA Ann. Conf. Atlantic City (June 29, 1978).
6. Vondrick, A.F. If It's Wet, They Use It in Phoenix Dual Distribution Systems. Proc. AWWA Dual Distribution Seminar. Denver, Colo. (June 26, 1976).
7. Coffelt, L.C. Information provided to AWWA Dual Distribution Committee.
8. Felke, D. Personal communication to L.A. Dove (May 22, 1978).
9. Miliken, T.G. & Trumbly, A.S. Municipal Recycling of Wastewater. *Jour. AWWA*, 71:10:548 (Oct. 1979).
10. *European Research Index* (2 Vols.). F.H. Books, Ltd., Guernsey, Channel Islands (3rd ed., June 1973).
11. Zabelin, V.A. & Rogachev, G.P. Water Consumption and Provision of Moscow and the Moscow District in the Future. *Water Resources* (trans. from *Vondyne Resursy*), 1:5:687 (Sept.–Oct. 1974).
12. Muller, W.J. Reuse of Wastewater in the Federal Republic of Germany. Org. Econ. Coop. & Devel., Paris (1969).
13. Frontinus, S.J. *Water Supply of the City of Rome* (Clemens Herschel, translator). London (1913).
14. Abrahams, H.J. The Water Supply of Rome. *Jour. AWWA*, 67:12:663 (Dec. 1975).
15. Passino, R. Personal communication to W.J. Boegly Jr. (May 11, 1976).
16. Shenderuk, G.N. Effect of Economic Activity on the State of Water Resources in West Germany. *Water Resources* (trans. from *Vodyne Resursy*), 1:1:146 (Jan.–Feb. 1974).
17. Stander, G.J. & Funke, J.W. Conservation of Water by Reuse in South Africa. Water Reuse, AICHE Sym. Series, 78:63:1 (1967).
18. Laburn, R.J. Some Aspects of Water Management in the P.W.V. Metropolitan Region. *Water Resources Bull.*, 12:1:65 (Feb. 1976).
19. Ganneau, M. Les Egouts de Paris. *Travaux*, 374:227 (1966).
20. Feasibility of Utility Tunnels in Urban Areas. Special Rept. 39, Washington, D.C. (Feb. 1971).
21. Anonymous. Pipe Subways in British Cities and in Paris. *Engrg. News*, 57:11:280 (Mar. 14, 1907).
22. Miller, D.G. & Newsome, D.H. Conservation of Water by Reuse in the United

Kingdom. *Water Reuse*, AICHE Sym. Series, 78:63:13 (1976).

23. DEB, A.K. & IVES, K.J. Dual Water Supply. Presented at the ASCE Environ. Engrg. Div. Second Ann. Conf. on Res. Devel. & Design, Gainesville, Fla. (Jul. 21–23, 1975).
24. RYABYSHEV, M.G. Sources of Moscow's Water Supply. *Water Resources* (trans. from *Vodyne Resursy*), 1:2:361 (May–June 1974).
25. GRIFFITH, C.O. Conservation of Water by Reuse in Mexico. *Water Reuse*, AICHE Sym. Series, 78:63:37 (1967).
26. SCHMIDT, C.J.; BEARDSLEY, R.F.; & CLEMENTS, E.W. III. A Survey of Industrial Use of Municipal Wastewater. *Complete Waterreuse*. (1973).
27. OKUN, D.A. Planning for Water Reuse. *Jour. AWWA*, 65:10:617 (Oct. 1973).
28. OKUN, D.A. The Heirarchy of Water Quality. *Envir. Sci. & Technol.*, 2:9:672 (Sept. 1968).

Chapter 3

Dual System Modeling

Analysis of Water Demand

The amount of water used in urban systems varies from system to system, but averages approximately 160 gpcd (605 L/d per capita). Urban water demand can be classified in five categories—residential, commercial, industrial, public, and unaccounted-for. Residential water demand, generally the largest component, can be broken down further to interior and exterior use as summarized in Table 3-1. Total water demand is a sum of the categorized demands and has the percent breakdown indicated in Table 3-2 as a national average.

Various researchers have determined interior residential water demand (shown in Table 3-3) for the following six categories: toilet flushing, bathing, laundry, dishwashing, drinking and cooking, and miscellaneous. High-quality potable water is required for drinking, cooking, bathing, laundry, and dishwashing; about 11 percent of the total interior household demand. The remaining 89 percent of interior residential water use may not require water of potable quality.

Exterior residential water demand varies, depending on season and area. On an average daily basis, exterior use amounts to about 7 percent of residential demand in Pennsylvania and 44 percent in California, with the difference between summer and winter representing the exterior residential usage. Since high-quality water is not required for exterior residential uses, the demand could be satisfied using nonpotable water.

Depending on the extent of dual supply in a community, the potable water requirement may vary from 11 percent to 60 percent of total water demand, which is about 7 gpcd (26 L/d per capita) to 36 gpcd (136 L/d per capita). The rest of the household water uses may be substituted by nonpotable water. The actual quantity of nonpotable water needed depends on location and season.

Irrigation and industrial reuse. Users who could substitute reclaimed waste-

Table 3-1. Urban Water Demand Categories

Interior Residential	Exterior Residential	Industrial
Toilet flushing Bathing Laundry Dishwashing Drinking and cooking Miscellaneous	Irrigating Car washing Swimming pools Cleaning	Manufacturing Processing Cooling
Public	**Commercial**	**Unaccounted for**
Schools Prisons Public hospitals Public parks	Office buildings Hotels Restaurants Car washes Laundries Golf courses Cemeteries Shopping centers Retail businesses	Leakage and loss Fire hydrant usage Testing Flushing Meter under-registration

Table 3-2. Urban Water Demands as a Percentage of Average Daily Use

Source	Category of Use					Total flow
	Residential	Commercial	Industrial	Public	Unaccounted for	*gpcd (Lpcd)*
McPherson[1]	33	12	33	7	15	
California DWR[2]	68	10	18	4		
Linaweaver, Geyer, & Wolff[10]	50					160 (605)
Haney and Hamann[13]	43	19	25	13		
USPHS[18]	41	18	24	17		
Fair, Geyer & Okun[21]	33	13		7	17	150 (568)
Bostian[22]	46	17	25	12		150 (568)
Hirshleifer, DeHaven, & Milliman[23]	45	18	32	5		
Murray and Reeves[26]	38	32		30		166 (628)
US Water Resources Council[27]	46	18	23	13		157 (594)
Frey, Gamble, & Sauerlender[28]	49	12	21	18		166 (628)
AWWA[29]	42	18	22	18		179 (677)
Weston National Water Utility Survey	52	17	15	7	9	153 (579)
Pennsylvania Water Utility Survey	39	12	31	5	13	162 (613)
Average	40	15	25	5	15	160 (605)

Table 3-3. Interior Residential Water Usage Comparison as a Percentage of Average Daily Use

Source	Category of Use						Total flow
	Toilet Flushing	Bathing	Laundry	Dishwashing	Drinking and Cooking	Miscellaneous	*gpcd (Lpcd)*
McPherson[1]	42	27	17		8	6	
California DWR[2]	42	32	14	12			
ERCO[3]	39	34	14	6	5	2	64 (242)
Laak[4]	47*	21*	18*	9			41 (155)
Murawczyk & Ihrig[5]							62 (235)
Ligman[6]	41*	26	19*	10			45 (170)
Wallman[7]	27–45*	18–36*	18*	13			30–50* (113–189)
Howe et al[8]	45	30	20		5		
Bailey & Wallman[9]	39	34	14	11		2	64 (242)
Durford & Becker[12]	41	37	4	11		7	
Haney & Hamann[13]	39	32	14	11		4	
Bennett[14]	33*	24*	27*	16			44.5* (168)*
Siegrist, Witt, & Boyle[15]	22*	23*	25*	11*		19*	
Water Encyclopedia[16]	42	38	7	11		2	
Bostian[17]	27–45	22–36	18	13		2	
USPHS[18]	30	35	20	15			
Chanlett[19]	43	38	7	11		11	
University of Wisconsin[24]	40	30	15	10		5	50 (189)
Bailey et al[25]	24	32*	4*	12		3*	
Average	40	30	15	6	5	4	60 (227)

*Rural areas

water for quality freshwater are divided into the following categories: irrigation (agriculture, landscape, other), industrial (process, cooling, boiler feed), groundwater recharge, and other.

In 1979 wastewater reuse for irrigation and by industry in the United States was identified to be about 680 mgd (2574 ML/d)[32] and can be broken down as follows:

Agricultural irrigation	200 mgd (757 ML/d)
Landscape irrigation	34 mgd (129 ML/d)
Other irrigation	188 mgd (711 ML/d)
Industrial process	66 mgd (250 ML/d)
Industrial cooling	142 mgd (537 ML/d)
Boiler feed	6 mgd (23 ML/d)
Groundwater recharge	34 mgd (129 ML/d)
Other	10 mgd (38 ML/d)
Total	680 mgd (2574 ML/d)

The preceding existing wastewater reuse has been projected to the potential wastewater reuse in the year 2000 as follows:

Agricultural irrigation	177.8 bgd (673 GL/d)
Landscape irrigation	1.4 bgd (5 GL/d)
Steam electric	80.1 bgd (303 GL/d)
Industrial cooling	16.9 bgd (65 GL/d)
Industrial other	14.1 bgd (53 GL/d)
Total	290.3 bgd (1099 GL/d)

It has also been estimated that only 26.7 bgd (101 GL/d) of the municipal wastewater effluent will be available for reuse in the year 2000. From these estimates it appears that there is a potential for water reuse through dual systems.

Table 3-4 shows actual (1975) and estimated (2000) potential wastewater reuse and availability of water for reuse. Some of the factors that will influence reuse potential include geographical location, timing of wastewater discharges and requirements of potential users (e.g., reuse requirements for irrigation are seasonal whereas municipal wastewater discharges are relatively uniform), and the availability and cost of alternative supplies. Figure 3-1 shows the reduction in potable water use in St. Petersburg when the dual water system was begun in 1976.

In areas where quality water is not available or is available in limited quantity, or where there is an overall shortage of water, developing a long-term economic water management plan will be complex. It might be necessary to consider a dual distribution system having more than one source of water for potable and nonpotable uses.

Dual Distribution System Modeling

Computer modeling is a useful tool for studying and comparing dual water systems and conventional water systems. The WATMAN Computer Model,[33] developed under a grant from the National Science Foundation, provides a technical and economic evaluation of various conventional and dual distribution alternatives for long-term planning and management of urban water systems, and helps urban water and wastewater managers determine the cost effectiveness of dual systems compared to conventional systems.

Table 3-4. Estimated Wastewater Reuse and Availability[30] for the Year 2000—mgd (ML/d)

Water Resources Region	Water Withdrawals Capable of Wastewater Reuse		Wastewater Discharges Available for Reuse				Wastewater Recycle		Wastewater Reuse	
			Treated		Untreated		Actual	Estimated Potential	Actual	Estimated Potential
	1975	2000	1975	2000	1975	2000	1975	2000	1975	2000
New England	3 615	1 424	4 563	2 108	0	0	2 134	16 284	3	13
	(13 684)	(5 390)	(17 273)	(7 980)			(8 078)	(61 642)	(11)	(49)
Mid-Atlantic	13 757	8 009	16 261	10 036	0	0	8 680	58 131	116	224
	(52 076)	(30 317)	(61 555)	(37 990)			(32 857)	(220 050)	(439)	(848)
South Atlantic-Gulf	22 645	25 311	18 799	17 180	653	833	16 071	127 414	7	192
	(85 721)	(95 813)	(71 162)	(65 033)	(2 472)	(3 153)	(60 835)	(482 314)	(26)	(727)
Great Lakes	38 564	20 410	40 059	20 737	0	0	9 589	97 417	10	290
	(145 981)	(77 260)	(151 640)	(78 498)			(36 298)	(368 763)	(38)	(1 098)
Ohio	32 514	13 913	33 018	12 443	0	0	19 032	112 417	1	118
	(123 079)	(52 666)	(124 987)	(47 102)			(72 044)	(425 545)	(4)	(447)
Tennessee	7 030	5 475	7 076	4 880	0	0	2 732	29 199	0	41
	(26 611)	(20 725)	(26 786)	(18 473)			(10 342)	(110 530)		(155)
Upper Mississippi	10 311	5 359	11 151	5 093	0	0	8 155	66 139	3	48
	(39 031)	(20 286)	(42 211)	(19 279)			(30 870)	(250 363)	(11)	(182)
Lower Mississippi	14 249	24 405	9 000	18 132	722	561	10 453	52 982	0	108
	(53 938)	(92 383)	(34 069)	(68 637)	(2 733)	(2 124)	(39 569)	(200 599)		(409)
Souris-Red-Rainy	256	594	211	54	3	28	123	308	0	1
	(969)	(2 249)	(799)	(204)	(11)	(106)	(466)	(1 166)		(4)
Missouri	40 385	47 016	5 126	5 791	1 767	1 901	5 476	37 795	6	1 139
	(152 874)	(177 975)	(19 404)	(21 921)	(6 689)	(7 196)	(20 729)	(143 070)	(23)	(4 312)
Arkansas-White-Red	13 489	13 542	1 848	1 773	390	346	12 530	42 479	75	147
	(51 061)	(51 262)	(6 995)	(6 712)	(1 476)	(1 310)	(47 431)	(160 801)	(284)	(556)

Continued on Next Page.

Table 3-4. Continued

Water Resources Region	Water Withdrawals Capable of Wastewater Reuse		Wastewater Discharges Available for Reuse				Wastewater Recycle		Wastewater Reuse	
			Treated		Untreated		Actual	Estimated Potential	Actual	Estimated Potential
	1975	2000	1975	2000	1975	2000	1975	2000	1975	2000
Texas Gulf	16 704	14 355	3 452	3 659	283	171	24 448	136 241	52	430
	(63 232)	(54 340)	(13 067)	(13 851)	(1 071)	(647)	(92 546)	(515 728)	(197)	(1 628)
Rio Grande	6 588	5 771	278	308	240	174	2 469	4 124	20	370
	(24 938)	(21 846)	(1 052)	(1 166)	(908)	(659)	(9 346)	(15 611)	(76)	(1 401)
Upper Colorado	7 564	8 251	223	353	598	559	2 650	7 598	1	61
	(28 633)	(31 233)	(844)	(1 336)	(2 264)	(2 116)	(10 031)	(28 762)	(4)	(231)
Lower Colorado	8 710	7 272	354	522	545	352	2 766	8 815	182	733
	(32 971)	(27 528)	(1 340)	(1 976)	(2 063)	(1 332)	(10 470)	(33 368)	(689)	(2 775)
Great Basin	8 280	7 190	470	591	640	454	1 586	11 511	4	62
	(31 343)	(27 217)	(1 779)	(2 237)	(2 423)	(1 719)	(6 004)	(43 574)	(15)	(235)
Pacific Northwest	45 322	40 172	3 528	1 902	4 563	3 438	3 677	28 288	26	67
	(171 562)	(152 068)	(13 355)	(7 200)	(17 273)	(13 014)	(13 919)	(107 082)	(98)	(254)
California	38 644	39 412	2 647	3 089	3 147	2 590	5 807	23 827	172	696
	(146 283)	(149 191)	(10 020)	(11 693)	(11 913)	(9 804)	(21 982)	(90 195)	(651)	(2 635)
Alaska	196	592	234	295	0	0	438	3 704	0	1
	(742)	(2 241)	(886)	(1 117)			(1 658)	(14 021)		(4)
Hawaii	1 813	1 173	300	205	178	88	197	808	1	7
	(6 863)	(4 440)	(1 136)	(776)	(674)	(333)	(746)	(3 059)	(4)	(26)
Carribean	619	616	318	452	0	0	29	44	1	5
	(2 343)	(2 332)	(1 204)	(1 711)			(110)	(167)	(4)	(19)
Totals	331 255	290 262	158 916	109 603	13 718	11 505	139 042	865 525	680	4 754
	(1 253 937)	(1 098 761)	(601 563)	(414 893)	(51 928)	(43 551)	(526 331)	(3 276 369)	(2 574)	(17 996)

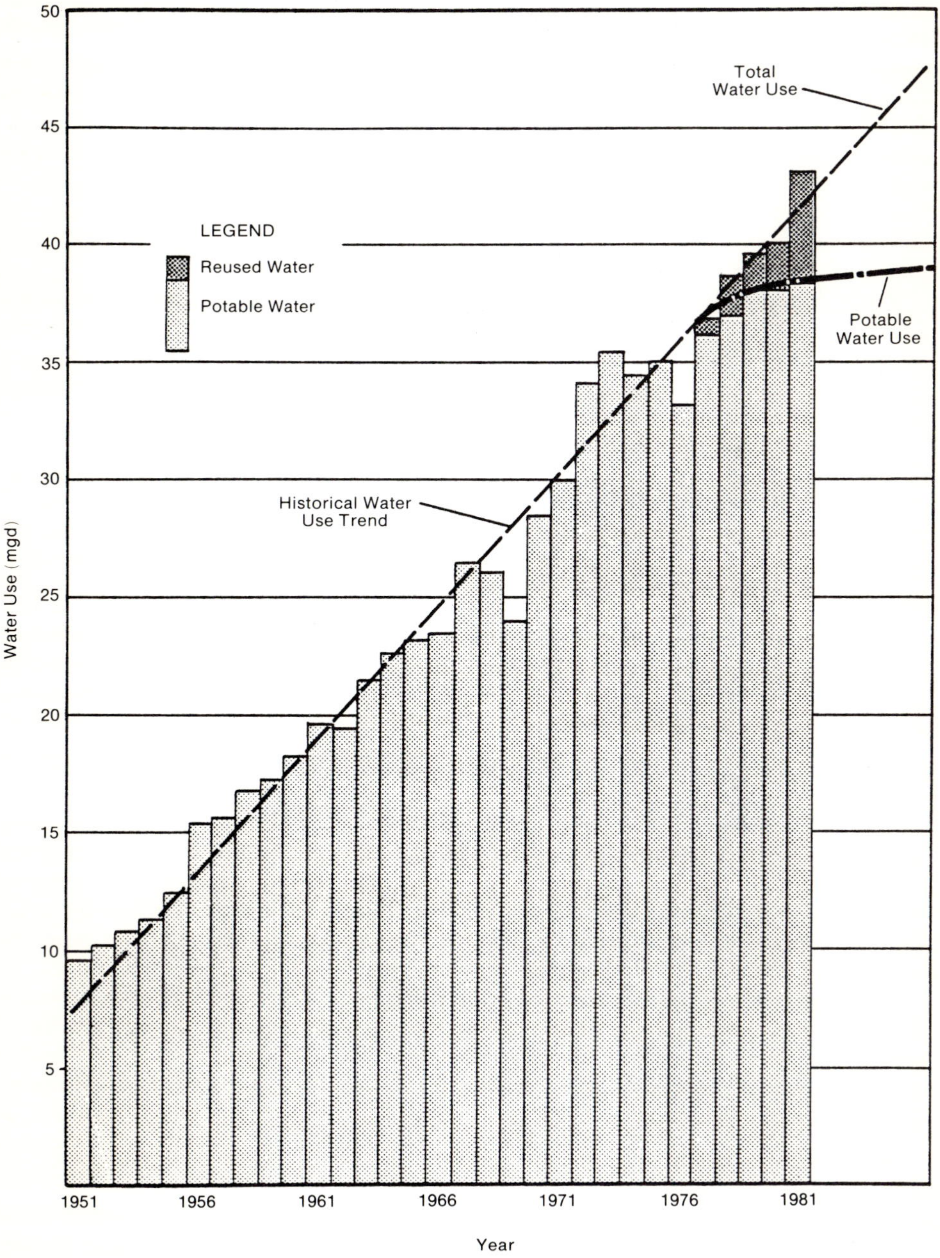

Figure 3-1. St. Petersburg record of water use.

A conceptual dual-supply urban water system model is shown in Figure 3-2. The model outline may have numerous options, depending on the quality and quantity of sources, potable and nonpotable water demand, and institutional and political considerations. This model is intended to provide a method of physical and economical analysis of urban water systems. It also provides the user with a tool to analyze various water supply system alternatives, including multiple distribution, depending on the composition of sources and demands. At present, this model can only be used as a planning tool rather than as a design tool.

This model has the following limitations: three qualities of water supplies, 10 communities or cities in a region, 50 unit processes or elements, and a 30-year planning period.

Potable and nonpotable segments of a dual distribution system modeled for analysis may be completely or partially separate, depending on the sources and treatment requirements. Total present worth of collection, transmission, treatment, and distribution of a water system of any configuration for single, dual, or multiple supply can be calculated using this model.

This method can be used for any urban water supply management and planning effort where a great number of alternatives are to be analyzed. This system model can specifically be applied to evaluate various alternative plans using conventional, dual, and multiple supply concepts in areas where the quality of the raw water source is poor, the availability of high-quality water is limited, or where there is an overall shortage of water.

New systems. In the design of new water supply systems, it is advisable to consider the multiple or dual supply concept as a viable alternative for a long-term safe water supply plan. From social, engineering, economical, and institutional points of view, it would be easier to establish a multiple or dual supply in a new city. Depending on the source of raw water, there may be many cases where a dual or multiple supply can be economically adopted.

Existing systems. With the introduction of the NIPDWRs, many existing water systems have found it necessary to provide further treatment in order to meet requirements. In cases where the cost of additional treatment for the bulk water is high, it might be appropriate to consider a dual supply system, using either the same source or an alternate source for potable supply. Figures 3-3 and 3-4 describe typical existing systems and outline conventional or dual supply alternatives. In cases where existing source quality is poor, it might be less expensive to treat a small fraction of the water to meet drinking water regulations and distribute it through a separate system. An alternative would be to introduce dual distribution systems gradually into existing cities as large users of nonpotable water are identified and as new areas are developed.

Water reuse systems. Common examples of reuse of wastewater for conservation of good quality sources are shown in Figures 3-5 and 3-6.

In an industrial city, where a large portion of the municipal water demand is for industrial needs, a dual supply can be effectively used with one nonpotable supply for industries and a potable supply for other uses. The industrial nonpotable supply can be derived from the same source with only existing treatment, or from reclaimed municipal wastewater with further treatment. Figure 3-7 shows a simple

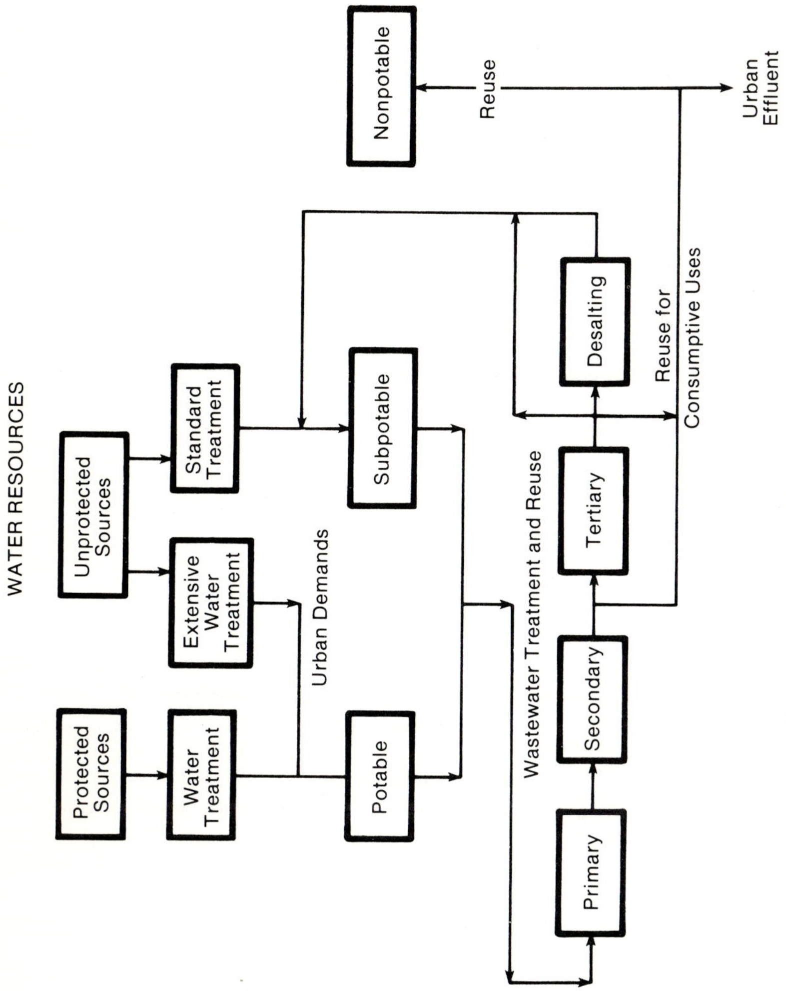

Figure 3-2. Dual supply urban water system.

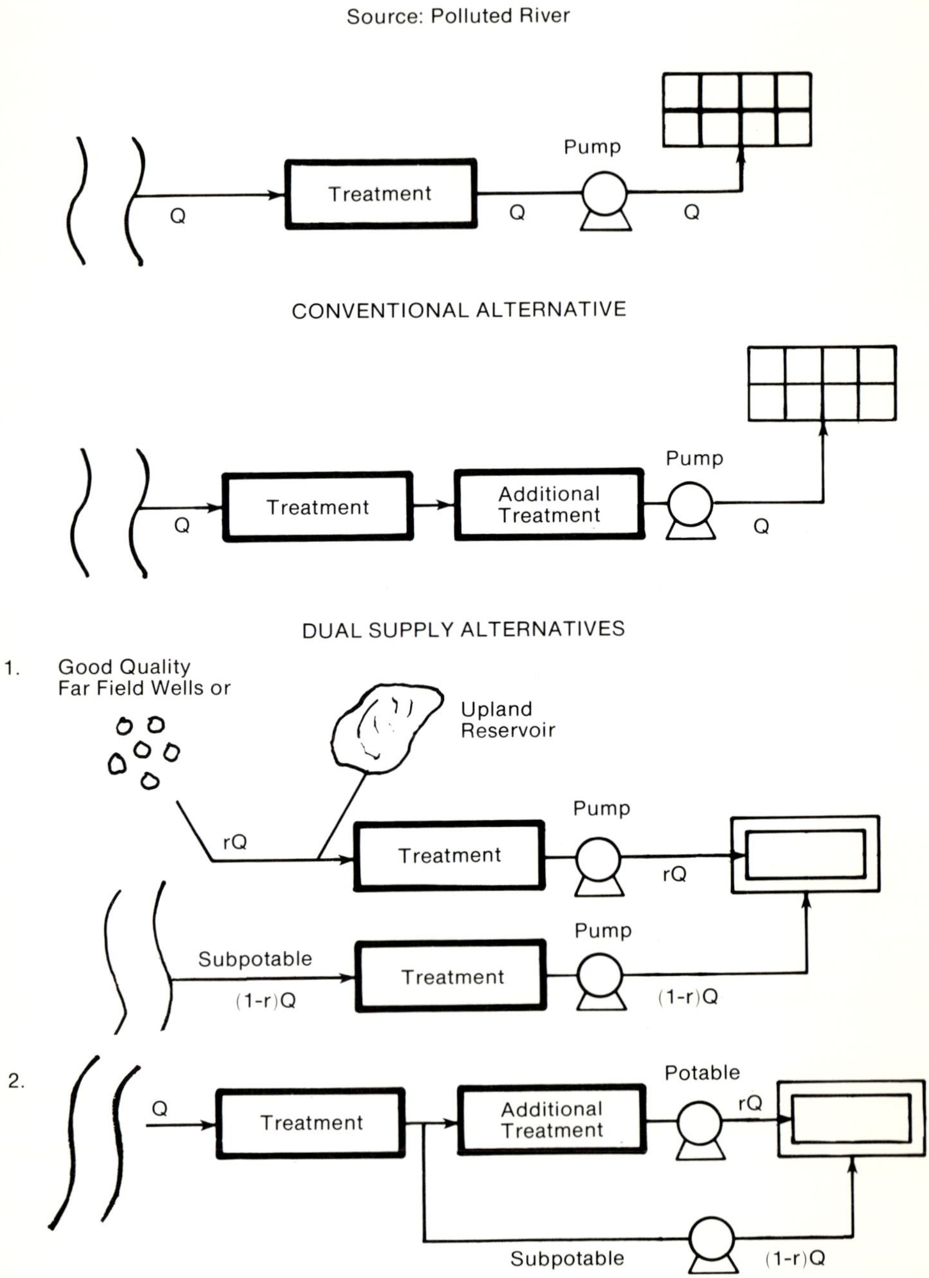

Figure 3-3. Existing system 1.

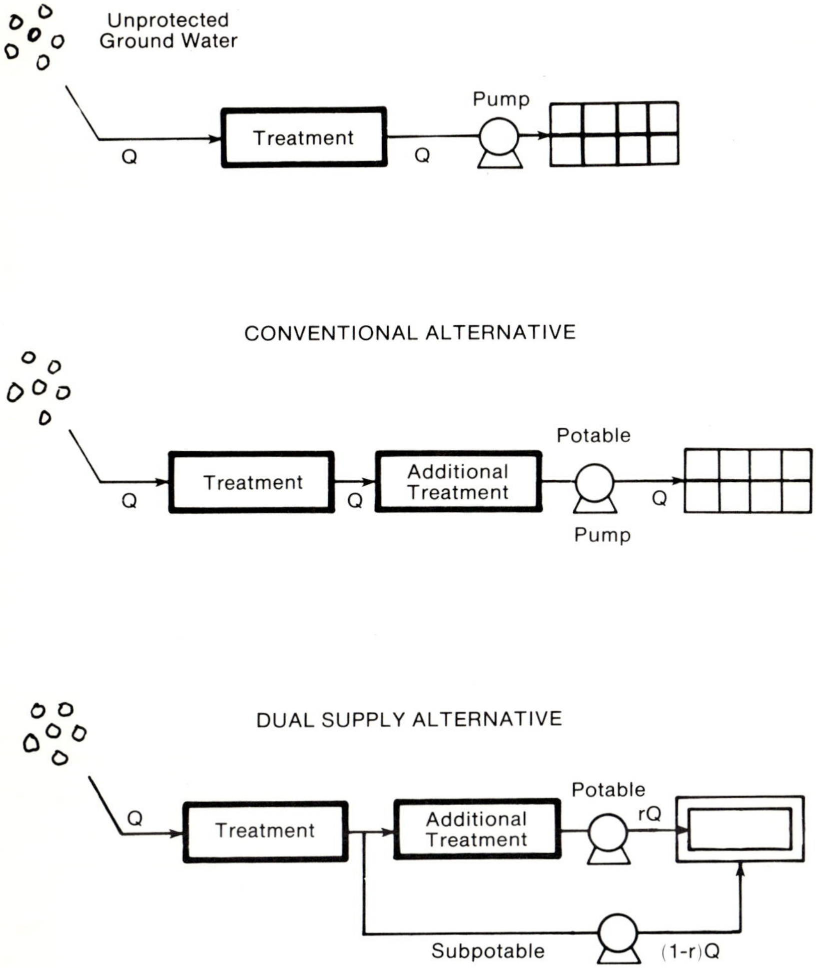

Figure 3-4. Existing system 2.

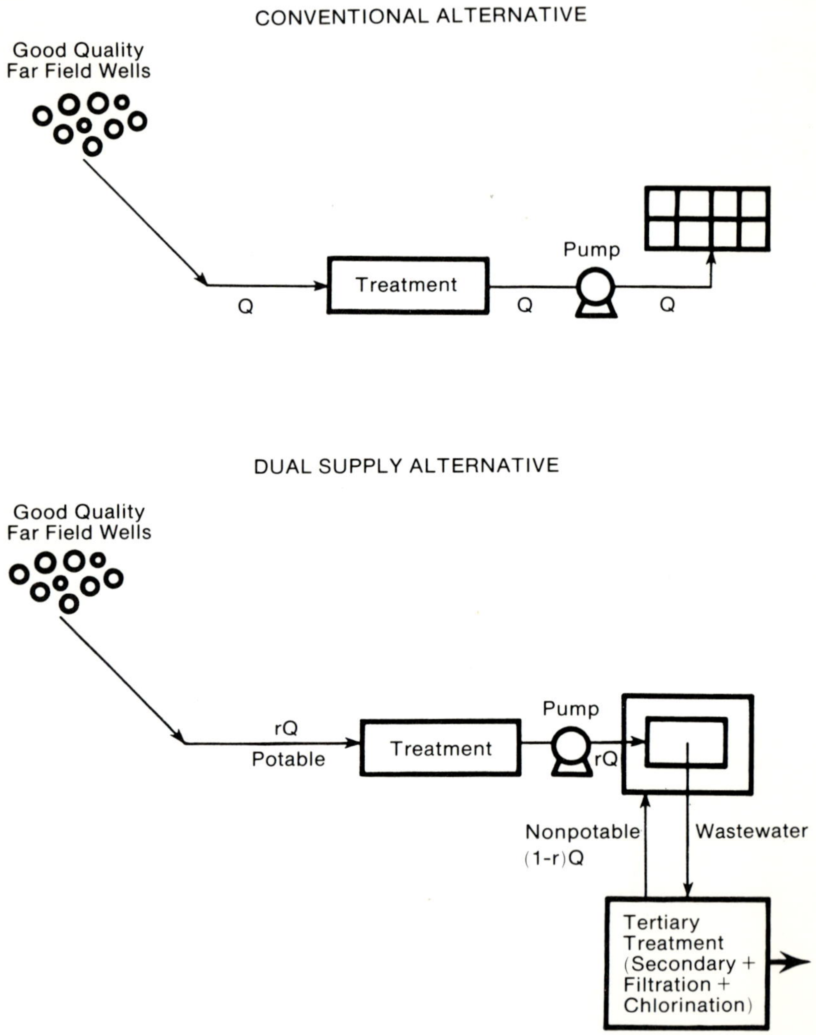

Figure 3-5. System 1-2 city: 100 000 population.

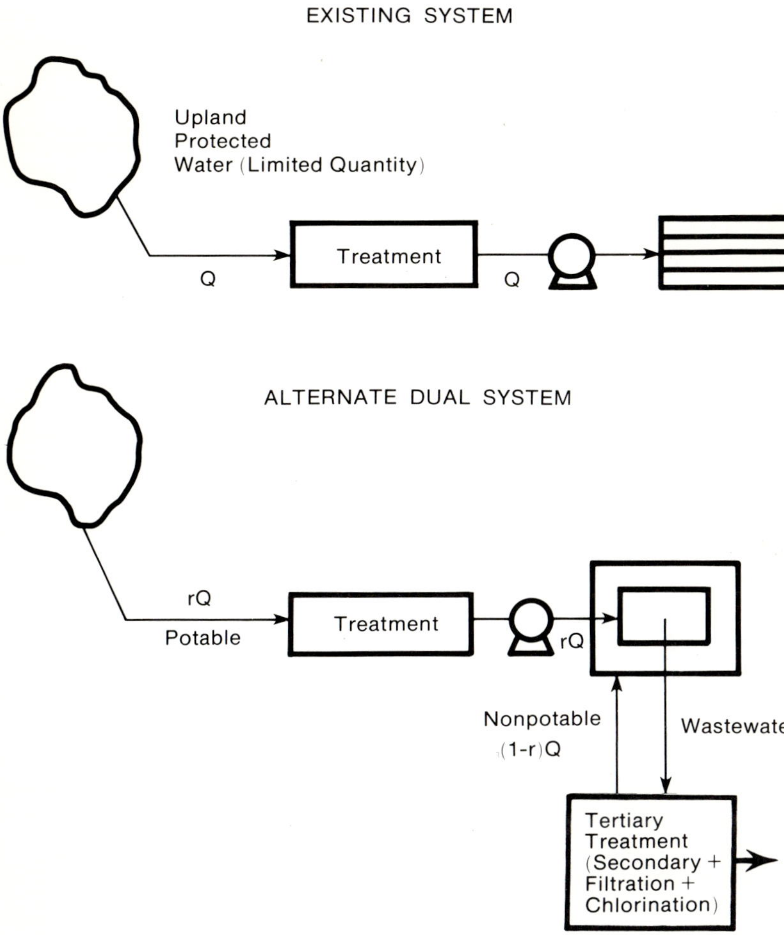

Figure 3-6. Existing system compared to alternate dual system.

schematic of alternatives for an existing system on a polluted river that has been analyzed with this model.

Regional systems. In regional management of water, high-quality limited sources can be conserved and used for potable supply through a regional water treatment and distribution system. Water for nonpotable uses can be obtained from reclaimed water or local polluted sources (Figure 3-8).

In a region where a limited quantity of protected water is available, existing communities that obtain their water supply from nearby polluted sources will have a choice. They will have the options of additional treatment for the removal of trace chemicals, or of using a limited protected source for potable supply to all communities in the region using a dual distribution system. The system shown in Figure 3-8 has been analyzed using the model.

Hypothetical Studies

Two of the water systems shown in Figure 3-3 have been analyzed using the model for various potable to total-flow ratios. The program has been run using three cities of different population sizes (20 000; 100 000; and 500 000). The computer output includes input data verification, details of costs and present worth of capital, operation and maintenance and salvage values, and total present worth of each system.

The ratios of total costs of dual supply to conventional supply for two systems (population 100 000), and for various potable to total-flow ratios are presented in Figures 3-9 and 3-10.

The water source for the conventional system in model 1 (Figure 3-9) is an unprotected river. In order to produce potable water from this source, granular activated carbon (GAC) has been added with the standard treatment of coagulation and filtration. In the corresponding dual system the potable source is quality groundwater, and the unprotected river source furnishes nonpotable water.

Figure 3-9 shows that the dual water supply cost is less than 85 percent of that for a conventional supply in a new system. If, however, the potable well source for a dual supply were 30 mi (48 km) away, dual supply would be more expensive than conventional supply (i.e., whenever the potable to total-flow ratio exceeds 0.25).

In model 2, unprotected river water is the only source for either a conventional or a dual system. Activated carbon treatment has been considered along with the standard treatment to produce potable water in the dual and conventional systems, whereas only standard treatment is given to produce nonpotable water in the dual system. The result is presented in Figure 3-10. A conventional system will be more economical whenever the flow ratio exceeds 0.2.

Conclusion

Quality water sources are limited and should be protected and conserved for uses requiring potable water. Water for other uses can be obtained from lesser quality sources, including reused water.

A separate distribution system will be necessary for supplying recycled water for nonpotable uses. In fact, dual supply systems would be a necessary feature of water management in areas where wastewater reuse will be considered.

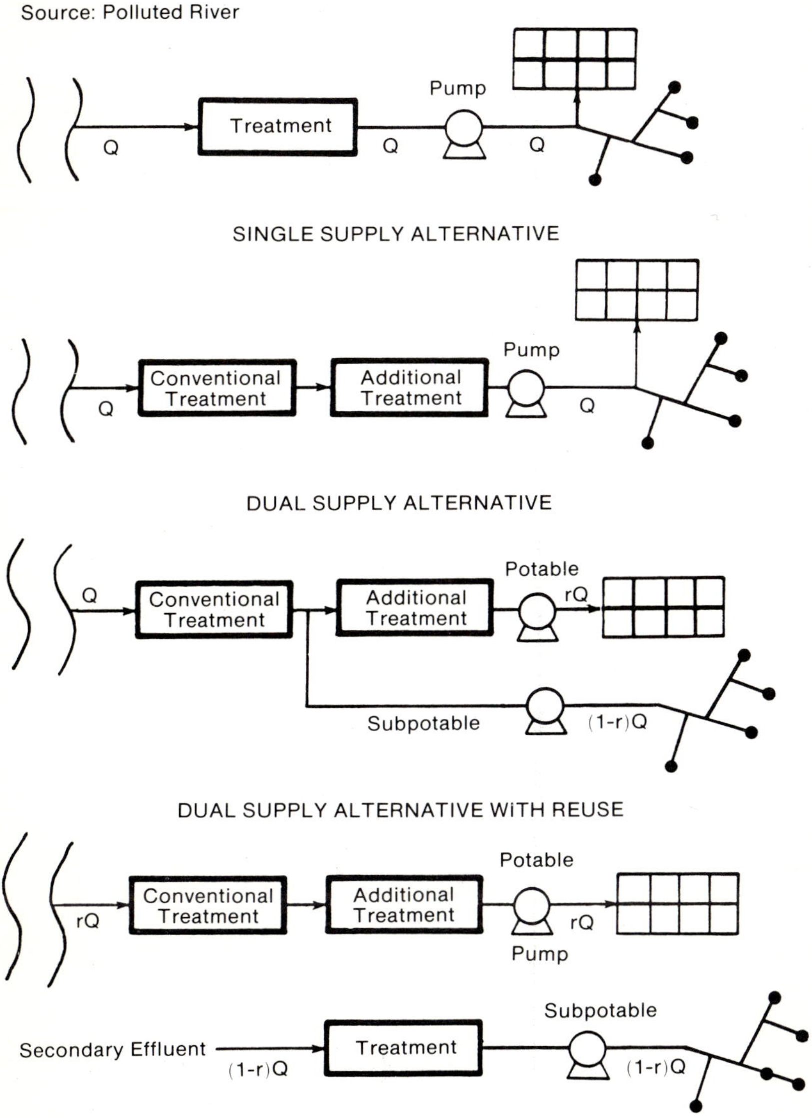

Figure 3-7. Existing system for industrial uses.

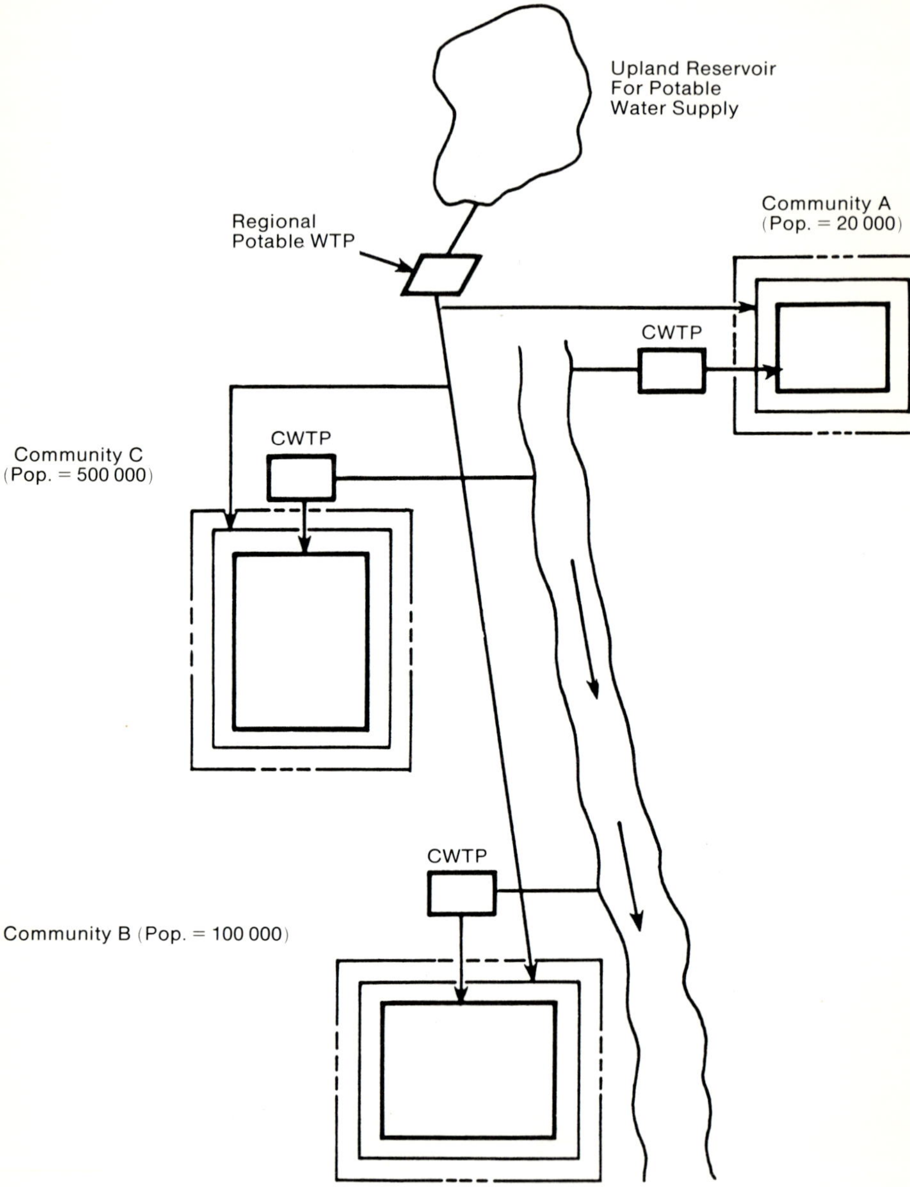

WTP = Water Treatment Plant

CWTP = Conventional Water Treatment Plant

Figure 3-8. Regional water supply management using dual supply concept.

CITY: 100 000 POPULATION

Conventional System

Source:	Unprotected River
Treatment:	Coag. + Sed. + Filt. + GAC + Disin.

Dual System

Potable Source:	Protected Ground Water
Treatment:	Disinfection
Subpotable Source:	Unprotected River
Treatment:	Coag. + Sed. + Filt. + Disin.

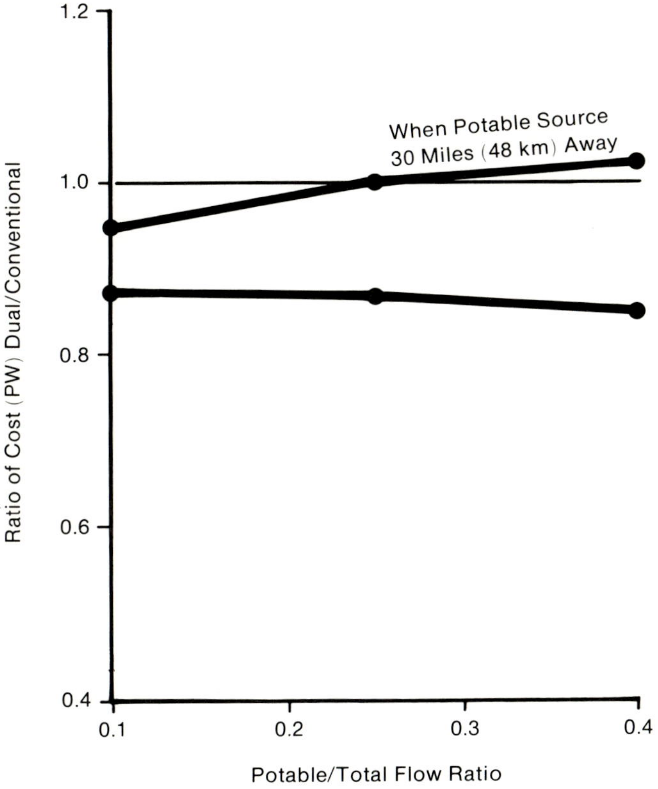

Figure 3-9. Dual water system model 1.

CITY: 100 000 POPULATION

Conventional System

Source:	Unprotected River
Treatment:	Coag. + Sed. + Filt. + GAC + Disin.

Dual System

Potable:	Unprotected River
Treatment:	Coag. + Sed. + Filt. + GAC + Disin.
Subpotable:	Same Source
Treatment:	Coag. + Sed. + Filt. + Chlorination

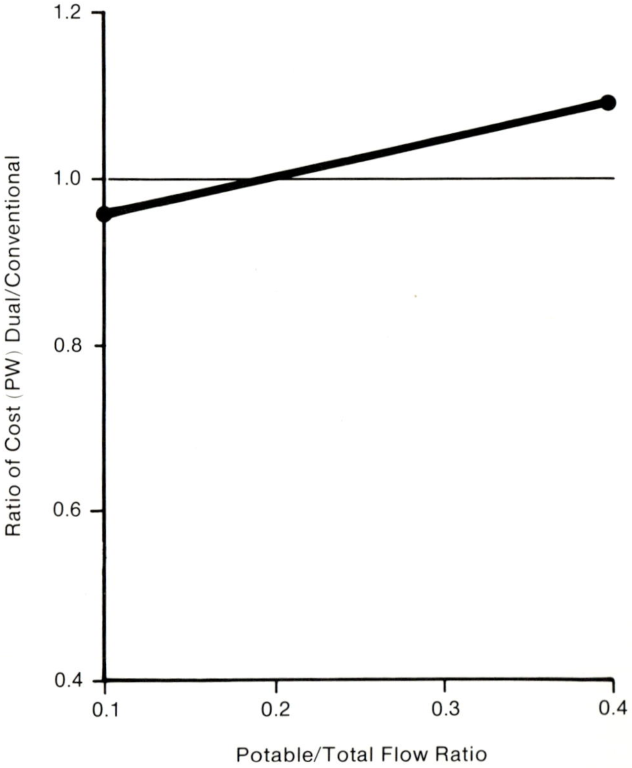

Figure 3-10. Dual water system model 2.

A dual water supply can also be adopted in cases where the raw water source quality is poor and has high organics, heavy metal, and total dissolved solids content. The cost of removing these materials is high and a dual supply system may be more economically viable than the conventional system. Again, this depends on the savings of additional treatment costs over the additional distribution costs.

A dual water supply should be an essential consideration in developing a long-term water management plan for a city or region considering conservation and optimum uses of water.

These test case results show that if GAC is required to produce potable water, a dual water supply will be cheaper than the conventional supply when the potable to total-flow ratio is less than 0.2.

References

1. McPherson, M.B. Conservation in Household Water Use. Presented at the Conf. on Planning Alternatives for Municipal Water Systems. French Lick, Ind. (Oct. 13, 1976).
2. California Department of Water Resources. Water Conservation in California. Bull. 198 (May 1976).
3. ERCO Energy Resources Company, Inc. Wastewater Flow Reduction Study. Final Rept. for USEPA, Contract No. 68-02-2964 (Oct. 1, 1975).
4. Laak, R. Relative Pollution Strengths of Undiluted Waste Materials Discharged in Households and the Dilution Waters Used for Each. Manual of Grey Water Treatment Practice—Part II. Monogmam Industries, Inc., Santa Monica, Calif. (1975).
5. Marawczyk, C. & Ihrig, D.C. Water Recovery and Solid Waste Processing for Aerospace and Domestic Applications. National Aeronautics and Space Admin., Manned Spacecraft Center, Houston (Jan. 1973).
6. Ligman, K. Rural Wastewater Simulation. Master's thesis, Univ. of Wisconsin, Madison (1972).
7. Wallman, H. Should We Recycle/Conserve Household Water? 6th Intern. Water Quality Sym., Washington, D.C. (Apr. 18–19, 1972).
8. Howe, C.W. et al. Future Water Demands—The Impacts of Technological Change, Public Policies, and Changing Market Conditions on the Water Use Patterns of Selected Sectors of the United States Economy: 1970–1990. Resources for the Future, Inc., Washington, D.C. (Mar. 1971).
9. Bailey, J. & Wallman, H. Flow-Reduction of Waste Water from Households. *Water & Sewage Works*, 118:3:69 (Mar. 1971).
10. Linaweaver, F.P. Jr.; Geyer, J.C.; & Wolff, J.B. Residential Water Use. Federal Housing Admin., Dept. of Housing and Urban Development, Washington, D.C. (June 1966).
11. Reid, G.W. Projection of Future Municipal Water Requirements. *Southwest Water Works*, S:46:18 (1965).
12. Durfor, C.N. & Becker, E. Public Water Supplies of the 100 Largest Cities in the United States, 1962. US Geological Survey Water Supply Paper 1812 (1964).
13. Haney, P.P. & Hamann, C.L. Dual Water Systems. *Jour. AWWA*, 50:9:1073 (Sept. 1965).
14. Bennett, E.R. & Linstedt, D.K. Individual Home Wastewater Characterization and Treatment. Completion Rept. Series 66, Environ. Resources Ctr., Colorado State Univ., Fort Collins (July 1975).
15. Siegrist, R.; Witt, M.; & Boyle, W.C. Characteristics of Rural Household Wastewater. *Jour. Environ. Engrg. Div.—ASCE*, 102:EE3:12200 (June 1976).
16. *The Water Encyclopedia* (David Todd, editor). Water Information Center, Port Washington, N.Y. (1970).
17. Bostian, H.E. News of National Environmental Research in Cincinnati. US Environmental Protection Agency, Cincinnati (Nov. 9, 1973).
18. US Public Health Service. *Manual of Septic Tank Practice*, Publ. 526, US Govt. Printing Ofce., Washington, D.C. (1967).
19. Chanlett, E.T. *Environmental Protection*, McGraw-Hill Book Co., New York (1973).
20. *CRC Handbook of Environmental Control, Volume VIII: Water Supply and Treatment* (R.G. Bond, C.P. Straub, and R. Prober, editors). Cleveland.
21. Fair, G.H.; Geyer, J.C.; & Okun, D.A. *Water and Wastewater Engineering*, Vol. 1, John Wiley and Sons, Inc., New York (1968).
22. Bostian, H.E. Water Conservation by the User. *APWA Reporter* (July 1974).

23. HIRSHLEIFER, J.; DEHAVEN, J.C.; & MILLIMAN, J.W. *Water Supply Economics, Technology and Policy*. Univ. of Chicago Press (1960).
24. Univ. of Wisconsin. On-Site Waste Water Disposal for Homes in Unsewered Areas. Small Scale Waste Management Proj. R2533 (Sept. 1973).
25. BAILEY, J.R. ET AL A Study of Flow Reduction and Treatment of Waste Waters from Households. General Dynamics, Electric Boat Division, Groton, Conn. (1969).
26. MURRAY, R.C. & REEVES, E.B. Estimated Use of Water in the United States in 1970. US Geological Survey Circular 676 (1972).
27. US Army Corps of Engineers, Omaha Urban Study (1976).
28. FREY, J.C.; GAMBLE, H.B.; & SAUERLENDER, O.H. Economics of Water Supply Planning and Management. Res. Publ. 90, Inst. for Res. on Land and Water Resources, Pennsylvania State Univ. (June 1975).
29. American Water Works Association. Operating Data for Water Utilities 1970 and 1965. AWWA Statistical Rept 20112, New York (1975).
30. WESNER, G.M. & HUGHES, M.V. JR. The Potential for Wastewater Reuse in the United States. Proc. Water Reuse Symp., Washington, D.C. (March 1979).
31. DEB, A.K. Multiple Water Supply Approach for Urban Water Management. Report submitted to the National Science Foundation, NSF ENV 76-18499 (1978).

Chapter 4

Dual Distribution System Design and Safeguards

Cost Effectiveness

The cost effectiveness of a complete dual system, based on conceptual design, is presented in Chapter 3 and again later in this chapter. The cost effectiveness of the limited dual water system constructed in the city of St. Petersburg, Fla., provides a useful case study.[1]

St. Petersburg implemented a limited dual system to conserve its water, financial, and energy resources. The public utilities director claims that, "Assuming that only half the treated wastewater is sold to commercial customers, the system should pay for its operating costs."[2] Such systems may be considered where water supplies are limited or distant, where advanced wastewater treatment (AWT) is required, or be based on a combination of these factors. Any cost-effectiveness analysis of a dual water supply system should consider all aspects of resource management.

St. Petersburg is in south Pinellas County, located midway down the west coast of the Florida peninsula. The county itself is shaped much like the Florida peninsula and lies between Tampa Bay and the Gulf of Mexico. It is nearly surrounded by salt water, and is in a water-short region.

Local water supplies were exhausted in 1928 when saltwater intrusion occurred in local wells pumping 2 mgd (7 ML/d). An engineering study concluded that the most cost-effective source of potable water would be a well field development in northwest Hillsborough County, 30 mi (48 km) away. That portion of the neighboring county was largely undeveloped. The first well field was expanded in the 1950s and a second one was constructed approximately 5 mi (8 km) farther away in the early 1960s. A third well field located another 4 mi (6 km) north in Pasco County was installed in the early 1970s, and the latest well field is located another 10 mi (16 km) distant. Water from the latest well field is pumped 50 mi (80 km) to the city for distribution.

The water supply facilities, well fields, pipelines, water treatment plant, and pumping stations are designed for 82 mgd (310 ML/d). The projected maximum daily water supply required for the year 2020 is 95 mgd (360 ML/d), or 13 mgd (49 ML/d) greater than the design of current facilities. Modifications to the facilities can be constructed for $10.5 million to force the additional water through the pipelines. A new pipeline and supporting facilities can be constructed for about $50 million, based on 1979 dollars.

Similar water supply costs will likely face any city that must import its high-quality water over long distances or that has an inadequate supply of high-quality water. Potable water does not have to be used for irrigation or many industrial purposes. When evaluating a dual water supply system, the savings to the potable system must be considered.

A desirable solution to the water supply problems facing St. Petersburg would be to reduce its ultimate potable water demand on the regional system. A second water system was designed to irrigate the massive institutional green spaces of the St. Petersburg metropolitan area and to serve residential lawn sprinkling, industries, and part of the firefighting demand. A nonpotable water system using highly treated wastewater as its supply source was selected.

The St. Petersburg wastewater treatment plant capital costs are compared to AWT in Table 4-1. St. Petersburg has four regional plants that range from 6.4-mgd (24.2-ML/d) to 20.0-mgd (75.7-ML/d) average daily flow. Florida Department of Environmental Regulation (FDER) guidelines for cost estimating show that standard secondary treatment plants to meet the total 62.4-mgd (236.2 ML/d) flow would cost $46.4 million. AWT plants would cost $78.1 million.

However, the dual water system approach added $6.6 million for filters, $19.7 million for 80 mi (129 km) of the dual distribution pipelines, $5.3 million for irrigation systems for public parks and schools, plus $4.9 million for deep-well storage in actual costs. Therefore, the total reuse concept totals $82.9 million, including the $5.3 million irrigation systems, compared to the AWT $78.1 million, excluding any irrigation systems. The capital cost based on the reuse concept, including irrigation systems and deep-well storage, is only 6 percent higher than the AWT alternative.

Operation and maintenance (O&M) costs for Florida wastewater treatment plants are examined in Table 4-2. The projected O&M costs for secondary and AWT plants were obtained from the FDER guidelines and reflect a secondary O&M cost of $2.272 million per year, compared to AWT at $6.032 million per year for the 62.4-mgd (236.2 ML/d) combined plant capacity. The O&M costs strictly associated with the reuse or dual water system amounts to $0.683 million per year. Therefore, an annual O&M savings of $3.077 million is realized by the city. These figures include debt service.

The treated wastewater is clear and odorless, with virus at a nondetectable level. It contains dissolved nitrates, phosphates, and potassium. These are of little health significance, but experience at one golf course in St. Petersburg indicated that an annual savings of 30 tons (27 200 kg) of fertilizer could be achieved on the 60 acres (24 ha) of fairways. This represents a 50 percent savings amounting to $6000. These

Table 4-1. St. Petersburg, Fla., Wastewater Treatment Plant Costs

Capacity of Existing Plants *gpcd (Lpcd)*	Capital Costs—*million dollars*		Actual Costs for Total Reuse* Concept—*million dollars*
	Secondary	AWT	
6.4 (24.2)	6.4	10.1	6.6 (filters†)
16.0 (60.6)	12.0	20.0	19.7 (distribution‡)
20.0 (75.7)	14.0	24.0	5.3 (irrigation§)
20.0 (75.7)	14.0	24.0	4.9 (deep wells)

*Costs for secondary and advanced wastewater treatment are based on Florida DER cost-estimating guidelines.
†Total reuse—46.4 (total secondary costs) + 36.5 (reuse costs) = 82.9
‡Cost for 80-mi (128.7-km) distribution system
§Cost for irrigation of 7600 acres (13 075 ha)

Table 4-2. St. Petersburg, Fla., Wastewater Treatment Plant Costs March 1979

Capacity of Existing Plants *mgd (ML/d)*	O & M Costs—*million dollars*		
	Secondary*	Reuse†	AWT*
6.4 (24.2)	0.374	0.070	0.86
16.0 (60.6)	0.584	0.175	1.51
20.0 (75.7)	0.657	0.219	1.82
20.0 (75.7)	0.657	0.219	1.82

*Secondary and advanced wastewater treatment plant costs are based on Florida DER cost-estimating guidelines.
†Total reuse—2.272 (total secondary costs) + 0.683 (reuse costs) = 2.955

results are based on an application rate of slightly more than 1 in. (25.4 mm) of irrigation water per week.

The US Soil Conservation Service (USSCS) recommends the annual nutrient application rate specified in Table 4-3. A typical homeowner would have a fertilizer bill of $12.44. It is doubtful that many homeowners apply fertilizer at this recommended rate. The nutrients supplied by slightly more than 1 in. (25 mm) per week irrigation is compared to the recommended rate and reflects $7.09 worth of value.

A cost comparison for evaluating the dual water supply system (reuse) and alternative AWT is presented in Table 4-4 for the total resource management system. The capital cost of the reuse system, including $5.3 million for irrigation systems, is $82.9 million, with an associated O&M annual cost of $2.955 million. The alternative AWT program would cost $78.1 million for wastewater treatment plants, $6.03 million in O&M annual costs, and necessitate $10.5 million in potable water system construction and $0.015 million in shallow irrigation wells, which would be unnecessary with the dual system. The capital cost required by the AWT program is $88.615 million without irrigation systems, compared to $82.9 million for reuse with irrigation. The savings with a dual system is more than 6 percent.

Operation and maintenance annual costs, including the fertilizer benefit, show a considerable savings of the dual system ($2.955 million) over AWT ($6.610 million). Experience in St. Petersburg shows that dual systems deserve careful and complete analysis when reviewing the total water resource management program.

Table 4-3. Nutrient Application Rate Recommended by USSCS for 1979

Nutrient	Amount Required *lb (kg)*	Amount Supplied* *lb (kg)*
Nitrogen	22.4 (10.1)	12.70 (5.8)
Phosphorus	8.4 (3.8)	4.16 (1.9)
Potassium	8.4 (3.8)	9.56 (4.3)

*By irrigating with St. Petersburg treated wastewater

Table 4-4. St. Petersburg, Fla., Resource Management System March 1979

Type of System	Costs—*million dollars*			
	Capital		O&M/yr	
Reuse (dual water system)	82.9		2.955	
Advanced wastewater treatment		78.100		6.032
Water		10.500		0.500
Private wells		0.015		0.002
Fertilizer benefit		0		0.076
Total	82.9	88.615	2.955	6.610

Energy Considerations and Sale of Treated Wastewater

The cost-effectiveness analysis could have been terminated at this point. However, it is appropriate to explore it a bit further. Energy in Florida is obtained mostly from oil. The difference in O&M costs between the dual system and AWT option is in energy and chemicals. There is only a minor difference in labor costs since the labor forces compare quite closely for the two systems.

Thus, it is important to note that most of the $3.655 million annual savings in the dual system, as shown on Table 4-4, is in costs directly or indirectly related to the petrochemical industry, i.e., energy and chemicals. These costs are increasing dramatically and will add significantly to the future cost effectiveness of the dual system alternative.

Conceptual Design of a Complete Dual System

The WATMAN model, reviewed in Chapter 3, provides a technical and economic evaluation of various conventional and dual distribution alternatives of urban water systems. The conceptual design of a complete dual system is presented here, based on a study by Haney and Beatty.[3]

Design analyses. The use of complete dual systems would result in some changes in design concepts for distribution grids.

The rates of use and demand required for designing conventional systems can be obtained by analyzing records of water production, storage, and pumping. Average daily use establishes revenues required to operate, maintain, and improve the system. Maximum daily use determines the capacity of treated supply facilities. Pumping capacity storage requirements and system design are based on either maximum hourly demand or maximum daily use plus fire flow. Comparable operating data are not presently available for nonpotable systems. It will be several years must elapse before design parameters based on operating experience can be

Table 4-5. System Information for Seven Utilities

Utility	Population Served *thousands*	Area Served *sq mi (sq km)*	Average Daily Use *gpcd (Lpcd)*	Total Residential Sales *percent*
Bloomington, Ind.*	64	28 (73)	148 (560)	21
Colorado Springs, Colo.	135	50 (129)	282 (1067)	35
Lincoln, Neb.	150	50 (129)	193 (730)	55
Madison, Wis.	172	50 (129)	169 (640)	45
Topeka, Kan.	148	65 (168)	140 (530)	62
Waterloo, Iowa	77	58 (150)	190 (719)	35
Wichita, Kan.	275	75 (194)	138 (522)	45

*All Indiana University use under industrial sales

developed and accepted. In the meantime approximate parameters may be established by analyzing operating data for conventional systems.

Operating data. Seven midwestern utilities were selected to demonstrate the effect of adding nonpotable water systems. The primary criterion for selection was the ready availability of operating data for their last full year of record, generally 1974. System information is shown in Table 4-5 for these utilities. Economic models of both conventional and dual systems were constructed, using available data for each of the utilities.

Water demand in dual systems. An average daily use of 40 gpcd (151 L/day per capita) was developed for the design of the potable system. Maximum daily use is assumed to be 150 percent of average daily use, or 60 gpcd (227 L/day per capita). Maximum hourly demand, if estimated as 150 percent of maximum daily use, is 90 gpcd (340 L/day per capita).

The differences between the parameters for the total water demand of the conventional system and the potable water demand of the dual system represent the approximate rates of use and demand for the nonpotable system. Tables 4-6 through 4-8 present the design rates of use and demand of other parameters that would result if potable systems were added to the existing systems for the various utilities. The existing systems would supply nonpotable water. The rates of nonpotable use and demand were assumed to be equal to the differences between the corresponding parameters for the conventional and potable systems. The validity of such an assumption may be questionable, particularly under conditions of maximum demand.

Detailed information necessary to prepare realistic estimates is not available because of a lack of experience with complete dual systems. Nevertheless, cost appraisals have been ventured using the annual reports from the seven utilities. Main sizes had to be scaled down to reflect the allocation of use and demand between the two distribution systems.

Size reductions of each utility were based on the relative hydraulic capacities required for maximum hourly demands. Indicated diameters were adjusted to conform to commercial pipe sizes.

The transformation patterns were calibrated utilizing available mathematical models for the Waterloo, Iowa system. Computer analyses were based upon maximum hourly demands and inputs for the potable and nonpotable systems. The

Table 4-6. Average Daily Use for Maximum Service with Dual Systems for Seven Utilities

Utility	Potable System *gpcd (Lpcd)*	Nonpotable System *gpcd (Lpcd)*
Bloomington, Ind.	40 (150)	108 (409)
Colorado Springs, Colo.	40 (150)	242 (916)
Lincoln, Neb.	40 (150)	153 (579)
Madison, Wis.	40 (150)	129 (488)
Topeka, Kan.	40 (150)	100 (378)
Waterloo, Iowa	40 (150)	150 (568)
Wichita, Kan.	40 (150)	98 (370)

Table 4-7. Maximum Daily Use for Maximum Service with Dual Systems for Seven Utilities

	Maximum Daily Use		Maximum Daily Use to Average Daily Use	
Utility	Potable System *gpcd (Lpcd)*	Nonpotable System *gpcd (Lpcd)*	Potable System *percent*	Nonpotable System *percent*
Bloomington, Ind.	60 (227)	153 (579)	150	142
Colorado Springs, Colo.	60 (227)	614 (2324)	150	254
Lincoln, Neb.	60 (227)	434 (1643)	150	284
Madison, Wis.	60 (227)	214 (810)	150	166
Topeka, Kan.	60 (227)	235 (890)	150	235
Waterloo, Iowa	60 (227)	239 (905)	150	159
Wichita, Kan.	60 (227)	194 (734)	150	198

Table 4-8. Maximum Hourly Demand for Maximum Service with Dual Systems for Seven Utilities

	Maximum Hourly Demand		Maximum Hourly Demand to Average Daily Use	
Utility	Potable System *gpcd (Lpcd)*	Nonpotable System *gpcd (Lpcd)*	Potable System *percent*	Nonpotable System *percent*
Bloomington, Ind.	90 (340)	192 (727)	225	178
Colorado Springs, Colo.	90 (340)	1090 (4126)	225	450
Lincoln, Neb.	90 (340)	656 (2483)	225	429
Madison, Wis.	90 (340)	329 (1245)	225	255
Topeka, Kan.	90 (340)	276 (1045)	225	276
Waterloo, Iowa	90 (340)	274 (1037)	225	183
Wichita, Kan.	90 (340)	305 (1155)	225	311

models were modified by reducing main sizes as indicated in the transformation patterns. The calibrated transformation patterns were used for the cost appraisals.

The distribution grids for the potable system were scaled down from what the authors considered to be a standard conventional grid. This grid has 8-in. (200-mm) mains 0.25 mi (0.4 km) apart in each direction and includes 6-in. (150-mm) mains in the remaining streets. Each 1-sq mi (2.6-sq km) area is enclosed by 12-in. (300-mm) or larger transmission mains.

The standard grid for the potable system was selected on the basis of replacing

Table 4-9. Comparative Installation Costs for Seven Utilities—*dollars per capita, 1976 Costs*

Utility	Conventional System	Dual System: Potable System	Dual System: Nonpotable System	Total Cost of Dual System
Bloomington, Ind.	342	252	303	555
Colorado Springs, Colo.	939	624	995	1619
Lincoln, Neb.	495	373	543	916
Madison, Wis.	360	278	387	665
Topeka, Kan.	412	324	429	753
Waterloo, Iowa	487	353	493	846
Wichita, Kan.	454	254	481	735

8-in. (200-mm) mains with 4-in. (100-mm) mains and replacing 6-in. (150-mm) mains with 2-in. (50-mm) mains. This grid was used for cost appraisals.

Additional service and plumbing changes. Addition of another system necessitates installation of a second service for virtually all customers. For cost appraisal the number of second services was assumed to be the same as the number of existing services. In addition, plumbing would be changed in homes and other buildings in order to provide nonpotable water to sill taps and toilets. The average costs for such changes were based on interviews with local plumbing contractors in areas served by three of the utilities considered. These costs reflect the work required for typical installations, but do not necessarily cover unusually difficult installations, such as houses built on concrete ground slabs. The relatively high costs for such installations might limit the extension of nonpotable water service.

Comparative installation costs. Table 4-9 presents appraisals of probable installation costs for adding a second distribution system. Appraisals for systems are included for purposes of comparison. Costs for existing mains, services, and fire hydrants are included.

Cost appraisals for potable systems are based on adding a 40-gpcd (151-L/day per capita) system. The existing system could be used to deliver nonpotable water. Cost figures include installation of second services and required plumbing changes.

Cost appraisals for nonpotable systems are based on the addition of a system that would meet existing demands. The existing system could be utilized for delivering potable water. Cost appraisals include installation of second services and necessary plumbing changes. Costs have been included for transferring fire hydrants from existing mains to newly installed nonpotable mains.

Under most circumstances the economical alternative would be the addition of a potable system. The existing system would be served from the nonpotable supply, which could be derived from available sources, reclaimed wastewater, or a combination.

All cost appraisals are based on typical urban installations and reflect 1976 construction prices. Appraisals shown for conventional systems have no relationship to historical costs or book values generated over a period of several years.

The cost appraisals indicate the approximate level of expenditure required to provide a dual system. The costs for second services and plumbing changes represent a substantial part of the cost of adding a second system.

Service lines and meters. For a conventional system, the standard service-line meter arrangement would be provided. For dual systems, each house and building served would theoretically require two service lines and two meters.

Plumbing. For a single-family residence served by a dual system, plumbing would be provided initially or modified to conform to the following: hot and cold water for kitchen sink, laundry facilities, showers, bathtubs, and lavatories would be obtained from a plumbing system served by the potable supply; toilets, sill faucets, and yard hydrants would be connected to the nonpotable service.

Firefighting requirements. Fire flow plus maximum daily use is an alternative to maximum hourly demand as a measure of system capacity. Although requiring less pumping capacity than maximum hourly demand, this combination may be more critical for the distribution system. Fire flow may be required at any point in the system and can cause high localized rates of flow. Only the nonpotable system would supply fire flows.

The seven utilities are large enough that maximum hourly demands probably exceed maximum daily uses plus fire flows under present circumstances. Providing dual systems would have the effect of reducing design populations for the nonpotable systems. Consequently, firefighting—maximum daily use plus fire flow—probably would replace maximum hourly demand as the critical loading for nonpotable system design in the smaller utilities, such as Bloomington, Ind., and Waterloo, Iowa. The maximum hourly demand would continue to be critical loading in larger utilities, such as Colorado Springs, Colo., and Lincoln, Neb. Either loading could be critical for the other utilities, and both conditions should be investigated for proper system design.

Modification of design concepts. In general, currently accepted design concepts apply to dual systems with little or no modification other than lower values of use and demand.

Conclusions. The following conclusions were indicated by the Haney and Beatty study.

1. Where conventional treatment is sufficient to produce a high-quality water, there is a definite annual cost difference in favor of the conventional system.
2. Where expensive treatment is required, e.g., demineralization, complete dual systems appear to have a slight cost advantage, although the margin is too small to be considered truly significant. The results may be characterized as interesting but inconclusive.
3. A condition definitely favoring complete dual systems exists where a limited source of good-quality groundwater is available. Assuming that this is sufficient for potable service needs and that the nonpotable supply can be obtained from an abundant water supply source, the dual system may offer a definite cost advantage.
4. Most significantly, when developed on the basis of dual sources as well as dual distribution, a dual system offers a means of conserving a limited supply of superior quality water.

Design Summary and Safeguards

The St. Petersburg limited dual system design has been successfully implemented and is meeting the planners' expectations. The innovative technology includes virus inactivation in the wastewater treatment system and deep well injection of the treated wastewater during periods when supply exceeds demand. The design of the dual distribution system varied little from the design of the conventional potable distribution system.

The Haney and Beatty study of seven midwestern cities concluded that complete dual systems involving the distribution of two grades of water, both safe, offer a means of providing consumers with water of quality that is related to specific needs. Water of excellent quality is provided for drinking, cooking, and cleansing. Safe water of lower quality is provided for all other uses. Where a limited service of good-quality groundwater is available, the dual system is usually cost effective.

The St. Petersburg limited dual system and the conceptual complete dual systems by Haney and Beatty indicate that the engineering design should pose no serious problems to experienced designers of conventional systems. The quality of pipe materials, appurtenances, and construction methods should be similar.

Numerous safeguards for the nonpotable water system were described in the St. Petersburg dual system and the Haney and Beatty study.

A dual distribution system questionnaire by Worth,[4] to which 28 states responded, disclosed concern about dual systems and little action by the states. Most states have not developed guidelines for dual systems, nor have they developed effective cross-connection control programs. The acceptance of dual systems by state agencies will depend greatly on the safeguards included to cope with the following and other concerns:

1. Public health is the overriding concern.
2. Cross-connection control must be provided.
3. Improper use of nonpotable water must be strictly controlled.
4. There must be adequate space for installing the dual system.
5. The limitations and conditions for use of the nonpotable water must be well defined.
6. The nonpotable system must be clearly marked and identified.
7. There must be close surveillance and monitoring of the nonpotable system.
8. The responsibility for the nonpotable system must be clearly established.
9. Construction and design standards for dual systems must be established.
10. The dual systems must be physically separated similar to existing water and sewerage systems.
11. Safeguards to minimize the effects of human carelessness must be established.

Numerous alternatives need to be examined and tested before a decision to install a dual system can be made. It is neither a quick nor inexpensive solution to water supply and distribution problems, but it is becoming more attractive and feasible as the supply of quality groundwater continues to decline.

References

1. Duynslager, W.A. St. Petersburg, Fla., Implements Reuse. AWWA Water Reuse Symposium, Washington, D.C. (Feb. 1979).
2. Dove, L.A. From Wastewater to Resource in St. Petersburg, Fla. Resource, Recovery, and Conservation Magazine. Eisevier Scientific Publishing Co., Amsterdam-Printed in the Netherlands. (1976/1977).
3. Haney, P.D. & Beatty, F.K. Dual Water Systems Design. *Jour. AWWA*, 69:7:389 (Jul. 1977).
4. Worth, J.E. State Guidelines and Regulations. AWWA Dual Distribution Systems Seminar, New Orleans, La. (June 20, 1976).

Chapter 5

Proposed Construction and Installation Procedures

Introduction

This chapter will examine the differences between the conventional potable water distribution system and the dual water distribution system. The potable water system normally covers the entire service area and reaches every consumer in both limited and complete dual systems. The nonpotable water system normally covers selected parts of the service area in a limited dual system and the entire service area in a complete dual system. As Haney and Beatty pointed out, it is usually more cost effective to use an existing water system for nonpotable water and construct a new system for potable water when converting to a complete dual system.[1]

Materials and Construction Procedures

The pipe materials and appurtenances should be of similar quality for both distribution systems. They both must be capable of providing lasting and dependable service under conditions that may vary in the future. Some special appurtenances that are required when constructing dual systems are discussed later in this chapter.

The principal construction difference between the two systems is that special identification and appurtenances are required to eliminate or greatly reduce the possibility of cross connections, backflow, or the inadvertent use of nonpotable water for potable purposes.

The procedures for trenching, installing, and backfilling pipelines and appurtenances should be similar for both systems. The pipelines of the two systems should be separated as far as conditions permit, both horizontally and vertically, to minimize the chances of damage and mixing the two waters in the event of rupture or leakage.

Construction of Complete Dual Systems

As discussed in Chapter 2, there are no complete dual systems serving entire communities in the United States nor have guidelines for complete dual systems been established by state governments and the responsible federal agencies.

The USEPA's Guidelines for Water Reuse[2] includes the California operational standard described in the next paragraph, which has one of the most complete guidelines for nonpotable water facilities.

The State of California Department of Health Services has published detailed regulations for reliability in wastewater-reclamation facilities; the regulations include discussion of emergency procedures, power supply, frequency of bacteriological analysis, and requirements for disinfection. In addition, Articles 8, 9, and 10 of the state's Title 22 regulations provide an excellent outline of design and operational considerations covering alarms, power supply, redundancy features, emergency storage and disposal, and chemical supply, storage, and feed facilities.

The California Department of Health Services has also established operational guidelines designed to assure the reliability and safety of reclaimed water programs. The guidelines for "Use of Reclaimed Water for Irrigation and Impoundments and Worker Protection at Water Reclamation Use Areas" contain numerous references for constructing nonpotable distribution systems. They are presented as Figures 5-1 and 5-2. An example of construction specifications for nonpotable distribution systems is shown in Figure 5-3.

Safety in Conveyance and Distribution

The USEPA's Guidelines for Water Reuse lists the following nonpotable urban reuse systems that are currently in operation in the United States:

Location	*Date of Initial Operation*
Grand Canyon Village, Ariz.	1926
Colorado Springs, Colo.	1960
Irvine Ranch Water District, Calif.	1975
St. Petersburg, Fla.	1977

Several other systems are now in the planning stages, such as the Walnut Valley (Calif.) Water District and Las Virgenes (Calif.) Municipal Water District.

The St. Petersburg and Irvine districts have nonpotable distribution systems totaling 82 and 52 mi (132 km and 84 km) in length, respectively. The primary use is for urban irrigation.

Identification of Nonpotable Pipelines

The following pipeline identification methods for nonpotable water systems are currently being used in various cities.

Warning tape. The plastic warning tape on pipelines has a red field with the words CAUTION RECLAIMED WATERLINE in black. The overall width is 3 in. (75 mm).

A. General
   1. Reclaimed water shall meet the Regional Water Quality Control Board requirements and the requirements specified in the Wastewater Reclamation Criteria, established by the State of California Department of Health for health protection.
   2. The discharge should be confined to the area designated and approved for disposal and reuse.
   3. Maximum attainable separation of reclaimed water lines and domestic water lines should be practiced.
      a. The use area facilities must comply with the "Regulations Relating to Cross-Connections," Title 17, Chapter V, Sections 7583-7622, inclusive, California Administrative Code.
      b. Plans and specifications for the existing and proposed reclaimed water system and domestic water system shall be submitted to state and/or local health agencies for review and approval.
   4. All reclaimed water valves and outlets should be appropriately tagged to warn the public and employees that the water is not safe for drinking or direct contact.
   5. All piping, valves, and outlets should be color-coded or otherwise marked to differentiate reclaimed water from domestic or other water.
      a. Where feasible, differential piping materials should be used to facilitate water system identification.
   6. All reclaimed water valves, outlets, and sprinkler heads should be of a type that can only be operated by authorized personnel.
      a. Where hose bibbs are present on domestic and reclaimed water lines, differential sizes should be established to preclude the interchange of hoses.
   7. Adequate means of notification should be provided to inform the public that reclaimed water is being used. Such notification should include the posting of conspicuous warning signs with proper wording of sufficient size to be clearly read.
   8. Adequate measures should be taken to prevent the breeding of flies, mosquitoes, and other vectors of public health significance during the process of reuse.
   9. Operation of the use area facilities should not create odors, slimes, or unsightly deposits of sewage origin.
B. Spray Irrigation of Crops
   1. Irrigation should be controlled to minimize ponding of wastewater and runoff should be contained and properly disposed.
   2. Irrigation should be done so as to prevent contact by the public with the sprayed material and precautions should be taken to ensure that reclaimed water will not be sprayed on walkways, passing vehicles, buildings, domestic water facilities, or areas not under control of the user.
      a. The irrigated areas should be fenced where primary effluent is used.

Figure 5-1. State of California Department of Health Guidelines for Use of Reclaimed Water for Irrigation and Impoundments.

b. Windblown spray from the irrigation area should not reach areas accessible to the public.
3. Irrigated areas must be kept completely separated from domestic water wells and reservoirs.
4. Adequate time should be provided between the last irrigation and harvesting to allow the crops and soil to dry.
    a. Animals, especially milking animals, should not be allowed to graze on land irrigated with reclaimed water until it is thoroughly dry.
5. There should be no subsequent planting of produce on lands irrigated with primary effluent.

C. Surface Irrigation of Crops
1. Irrigation should be controlled to minimize ponding of wastewater and runoff should be contained and properly disposed.
2. The public should be effectively excluded from contact with the reclaimed water used for irrigation.
    a. The irrigated areas should be fenced where primary effluent is used.
3. Irrigated areas must be kept completely separated from domestic water wells and reservoirs.
4. Adequate time should be provided between the last irrigation and harvesting to allow the crops and soil to dry.
    a. Animals, especially milking animals, should not be allowed to graze on land irrigated with reclaimed water until it is thoroughly dry.
5. There should be no subsequent planting of produce on lands irrigated with primary effluent.
6. Adequate measures must be taken to prevent any direct contact between the edible portion of the crops and the reclaimed water.

D. Impoundments
1. Runoff should be contained and properly disposed.
2. At restricted recreational impoundments and landscape impoundments all valves and outlets should be appropriately tagged to warn the public that the water is not safe for drinking or bathing.
3. At nonrestricted recreational impoundments all valves and outlets should be appropriately tagged to warn the public that the water is reclaimed from sewage and is not safe for drinking.
4. Adequate measures should be taken to prevent body contact activities, such as wading or swimming, at restricted recreational impoundments containing reclaimed water.
5. Adequate measures should be taken to prevent direct public contact with reclaimed water at landscape impoundments.
6. Restricted and nonrestricted recreational impoundments should be maintained under the continuous supervision of qualified personnel during periods of use.
7. Impoundments containing reclaimed water must be completely separated from domestic water wells and reservoirs.

Figure 5-1. (Continued).

1. Employees should be made aware of the potential health hazards involved with contact or ingestion of reclaimed water.
2. Employees should be subjected to periodic medical examinations for intestinal diseases and to adequate immunization shots.
3. Adequate first aid kits should be available on location, and all cuts and abrasions should be treated promptly to prevent infection. A doctor should be consulted where infection is likely.
4. Precautionary measures should be taken to minimize direct contact of employees with reclaimed water.
   a. Employees should not be subjected to reclaimed water sprays.
   b. For work involving more than a casual contact with reclaimed water, employees should be provided with protective clothing.
   c. At crop irrigation sites, the crops and soil should be allowed to dry before harvesting by employees.
5. Provisions should be made for a supply of safe drinking water for employees. Where bottled water is used for drinking purposes, the water should be in contamination-proof containers and protected from contact with reclaimed water or dust.
   a. The water should be of a source approved by the local health authority.
6. Toilet and washing facilities should be provided.
7. Precautions should be taken to avoid contamination of food taken to areas irrigated with reclaimed water, and food should not be taken to areas still wet with reclaimed water.
8. Adequate means of notification should be provided to inform the employees that reclaimed water is being used. Such notification should include the posting of conspicuous warning signs with proper wording of sufficient size to be clearly read.
   a. In some locations, especially at crop irrigation use areas, it is advisable to have the signs in Spanish as well as English.
9. All reclaimed water valves, outlets, and/or sprinkler heads should be appropriately tagged to warn employees that the water is not safe for drinking or direct contact (direct contact is allowed at nonrestricted recreational impoundments).
10. All piping, valves, and outlets should be color-coded or otherwise marked to differentiate reclaimed water from domestic or other water.
    a. Where feasible, differential piping materials should be used to facilitate water system identification.
11. All reclaimed water valves, outlets, and sprinkler heads should be of a type that can only be operated by authorized personnel.
    a. Where hose bibbs are present on domestic and reclaimed water lines, differential sizes should be established to preclude the interchange of hoses.

Figure 5-2. State of California Department of Health Guidelines for Worker Protection of Water Reclamation Use Areas.

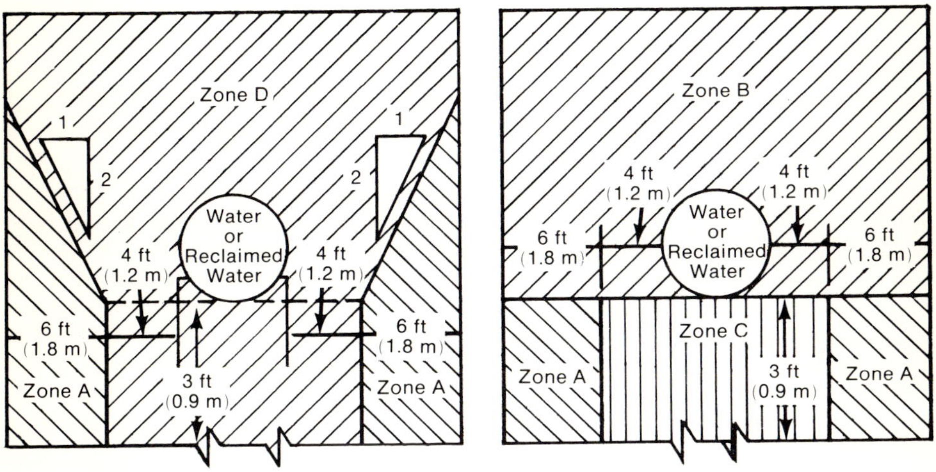

PARALLEL CONSTRUCTION PERPENDICULAR CONSTRUCTION

If any of the three pipelines—water, reclaimed water, or sewer—are within any of the above indicated zones, then special construction will be required as indicated below.

CONSTRUCTION REQUIREMENTS

Zone	Sewer	Reclaimed Water
A	VCP with compression joints.	ACP or PVC as specified in Part III.
B	CIP with lead or approved mechanical joints or VCP in steel casing.	CIP with lead or approved mechanical joints, or ACP, or PVC in steel casing.
C	VCP encased in concrete or any of Zone B construction methods.	ACP or PVC encased concrete or any of Zone B construction methods.
D	Do not locate any parallel sewer in this area without health department approval.	Do not locate any parallel reclaimed water main in this area without health department approval.

ACP with rubber ring joints approved for force mains may be placed in Zone A.

Mechanical compression joints conforming to ASTM C425, Vitrified Clay Pipe Joint Using Material Having Resilient Properties, examples are "Wedgelock" and "Speedseal" joints.

VCP = vitrified clay pipe; CIP = cast iron pipe; ACP = asbestos-cement pipe; PVC = polyvinylchloride pipe.

Figure 5-3. Example of construction specifications for dual distribution system.

Source: Irvine Ranch Water District. Standard Specifications for the Construction of Water, Sewer, and Reclaimed Water Facilities. Orange County, Calif., Mar. 5, 1979.

Warning tapes are installed directly on the top of the pipe longitudinally and are centered. The warning tape is continuous over the entire length of the pipe and is fastened to each pipe length by plastic adhesive tape bands around the pipe and warning tape at no more than 5-ft (1.5-m) intervals. All risers between the main line and control valves are installed with warning tape.

Stenciled pipe. Stenciled distribution system pipe is accepted as an alternative to warning tape. The words RECLAIMED WATER are stenciled in 2-in. (50-mm) green letters in two places on the pipe in at least three areas of a 13-ft (4-m) section of pipe for a total of six places per section of pipe.

Stenciled PVC irrigation pipe is white with green stenciling. The stenciling appears on both sides of the pipe with the marking RECLAIMED WATER in ⅝-in. (15-mm) letters repeated every 12 in. (300 mm).

Vinyl tape. All pipe 2.5 in. (64 mm) in diameter and larger, is identified by paper-backed adhesive vinyl tape. The tape is 6 in. (150 mm) wide with a brown background and yellow letters 1 in. (25 mm) high. The wording TREATED WASTEWATER MAIN is placed longitudinally on the pipe. There is a 1-ft (300-mm) separation between identifiers. For 2.5-in. (64-mm) through 4-in. (100-mm) pipe, the tape is centered along the top of the pipe; for 6-in. (150-mm) through 16-in. (400-mm) pipe, the tape is placed on both sides of the top half of the pipe. For 20 in. (500 mm) and larger pipe, the tape is placed as on the 6-in. (150-mm) through 16-in. (400-mm) pipe with a third strip along the center of the top of the pipe. The tape is placed from joint to joint.

Alternate pipe identification. Some contractors elect to use an alternate form of identification for treated wastewater pipes, such as painting, dyeing, stenciling, or coating the pipe either at the factory or in the field.

Identification and Description of Appurtenances

Several different methods are used to identify and safeguard appurtenances in nonpotable water systems, as follows.

Fire hydrants. Treated wastewater fire hydrants have a tamper-proof hold-down nut and can be operated only by a special wrench. They are painted dark brown with a yellow stripe, 12 in. (300 mm) wide around the barrel.

Valves. Valves are equipped with standard cast-iron roadway boxes. They are usually three-piece, slip type, with square cover. Valve box covers are cast with the words TREATED WASTEWATER.

Valve boxes are mounted in a concrete pad with a bronze disc giving the size, type of valve, and number of turns to open.

Control and regulating valves. All gate valves, manual control valves, electrical control valves, pressure-reducing valves, and pressure-relief valves for on-site reclaimed water systems are installed below grade in a valve box with a green cover.

Washdown hydrants and other points of public access. All on-site reclaimed water facilities are restricted from public access so that the general public cannot draw water from the system. Facilities such as washdown hydrants (typically found at tennis courts), blowoff hydrants, blowoffs on strainers, are restricted from public access.

These facilities, both above and below grade, are housed in an approved green lockable container. Signs reading CAUTION: RECLAIMED WATER are installed. An alternative means of restricting public access is the use of valves that operate by means of a recessed key slot or by means of hexagonal heads, such as those typically found on fire hydrants.

Warning signs and labels. Warning labels are installed on designated facilities such as controller panels and washdown or blowoff hydrants on water trucks and temporary construction services. The labels will notify that the system contains reclaimed water that is unsafe to drink.

Where reclaimed water is used for recreational impoundments, warning signs are installed to notify that the impoundment is unsafe for bodily contact.

Other safety practices. In addition, the following steps are taken in dual distribution systems.

1. Meters at spray irrigation sites are different from those used for potable water.
2. Meters may not be required on treated wastewater services to residential lots in highly developed areas.
3. Public education releases are distributed periodically to describe the system to new owners of homes and businesses.
4. An extensive monitoring program is maintained at treatment plants, deep wells, distribution systems, and sites where treated wastewater is used.
5. All treated wastewater storage ponds are covered to retard the formation of algae and to prevent dust and other airborne debris from contaminating the water and ultimately clogging the irrigation sprinklers.

There are many other safeguards that should be implemented in the construction and operation of nonpotable water systems. When constructing nonpotable urban irrigation systems, all connections to existing potable water supplies should be severed, such as to drinking fountains and lavatories. The affected items should be reconnected to a separate potable water system.

Sprinkler irrigation systems should have a Y strainer or basket strainer located immediately downstream of the meter. Strainers that have automatic backwash features should not normally be allowed unless it can be demonstrated that the backwash water will not cause runoff and is disposed in an approved manner. The strainer drain valve should operate with a recessed key slot.

Recreational or decorative bodies of water designed to utilize a water filter system must be designed to discharge the backwash in an approved manner.

Quick-coupling valves should be used in nonpotable water systems. The valves should be operated with a special coupler key with an acme thread for opening and closing. The cover should be attached permanently to the valve.

Backflow prevention devices should be installed on all potable water service connections entering private properties served by the nonpotable system. The device should consist of a reduced-pressure backflow prevention device or a double check-valve assembly, depending on the degree of potential hazard.

Special protection encasements should be placed around nonpotable water lines at all points where they cross over or under, or closely parallel potable water lines.

Summary and Recommendation

The construction, materials, and installation for potable and nonpotable pipelines and appurtenances in a dual water distribution system should be of similar quality. The nonpotable system that is designed for lower cost and lower quality may be unable to serve new applications in the future that require higher water pressure or other changed conditions.

The major concern in the construction and installation of dual systems is to identify and safeguard all pipelines and appurtenances to avoid their inadvertent cross connection, backflow, or direct use of nonpotable water.

References

1. HANEY, P.D. & BEATTY, F.K. Dual Water Systems Design. *Jour. AWWA*, 69:7:389 (Jul. 1977).
2. USEPA. Guidelines for Water Use. Natl. Tech. Information Service, EPA-600/8-80-036. Springfield, Va. (Aug. 1980).
3. Irvine Ranch Water Dist. Standard Specifications For the Construction of Water, Sewer and Reclaimed Water Facilities. Irvine, Calif.
4. Irvine Ranch Water Dist. Rules and Regulations For Water, Sewer, and Reclaimed Water Service. Irvine, Calif.
5. Rules and Regulations. For the installation, operating, and metering of the treated wastewater irrigation system. Amendment to Ordinance 309.F, City of St. Petersburg, Fla. (Oct. 19,1978).

Chapter 6

Management

The management of a dual water system must take into account all aspects of the system, including planning, design, construction, operation, maintenance, monitoring, and surveillance activities. Ideally, one agency would be responsible for managing the entire operation from treatment to use. In most cases, however, at least two different agencies will be responsible for control of the treatment, distribution, and use of the secondary supply. Therefore, prior to implementation of any dual system project, the agencies responsible for the operation and management of the system must be clearly defined, and the specific duties, responsibilities, and liabilities of each agency must be recognized and accepted by all affected agencies and regulatory authorities.

Planning should take place on a regional level in order to develop an integrated system that may supply water for several different types of use. Economic analyses will determine the size and complexity of a proposed system from a cost-effectiveness standpoint. Detailed financial information, including projected costs and sources of funds, will have to be developed and evaluated during the early planning stages.

It will be essential to provide an uninterrupted supply of high-quality water in order to ensure customer satisfaction. Otherwise, the customer may abandon the practice of utilizing two grades of water. Hence, the guidelines for the design, operation, and supervision of treatment and distribution facilities should be of the same order as are employed for potable water treatment and distribution.

For the purposes of this manual, "off-site facilities" means facilities under control of the purveyor of the secondary supply, including but not limited to treatment facilities, secondary water pipelines, reservoirs, pumping stations, fire hydrants, manholes, valves, connections, supply interties, and other appurtenances and property up to the point of connection with the customer's facilities. "On-site facilities" means facilities under the control of the user (applicant, owner, or customer), or those downstream of the service connection, which will normally be the downstream end of the meter tailpiece.

The following guidelines are based, in part, on those developed for the Irvine Ranch Water District, Orange County, Calif.

Service Conditions

a. A secondary water supply shall be provided only if a permit for such water service is obtained from the purveyor or state regulating agency. The permit and permitting procedure must be approved by the appropriate regulatory agencies.

b. No water shall be provided that does not comply with the treatment and quality requirements specified.

c. Water shall be available only in accordance with all applicable federal, state, and local statutes, ordinances, regulations, and contracts, as well as the terms of any service agreements and permits issued by the purveyor.

d. Water service may be terminated by the purveyor on a temporary or permanent basis at any time the customer's operations do not conform to the applicable requirements.

e. Each water service line shall be equipped with a curb stop or wheel valve on the inlet side of the meter, so that the purveyor can control the flow of water to the use area.

f. Each use area (irrigated land, residence, building) under separate ownership must be provided with a single and separate water service connection and water meter. Customers shall be held responsible and billed for all water passing through their meters. The nonpotable water services shall be metered in the same way that potable water services are metered, with reading of the meter and billing to be on the same schedule.

Backflow Prevention

a. The potable water supply must be protected (1) against actual or potential cross connections by isolating, within the premises, contamination or pollution that may occur because of some undiscovered or unauthorized cross connection on the premises, and (2) to prevent cross connection from occurring in the future.

b. Approved backflow prevention devices shall be installed at the water service connection to any premises where there is an auxiliary water supply carried in a separate piping system.

c. Backflow prevention devices should be inspected at least once each year. All inspections and testing shall be performed by a tester certified by the local health department. State or local regulatory agencies shall conduct independent inspections and institute cross-connection control programs as they deem necessary.

Water Supervisor

A water supervisor shall be designated at each location where the local health department or other regulatory authority deems it necessary. The water supervisor shall be responsible for the installation and use of pipelines and equipment and for the prevention of cross connections.

Authorized Uses for Secondary Water

The water purveyor may determine, at its discretion, whether it is necessary or desirable to furnish reclaimed water for the specific use involved. Determinations as to the acceptability of specific uses shall be made in accordance with all appropriate requirements and only after approval from the appropriate regulatory agencies.

Maintenance

a. Unless otherwise specified, the purveyor shall be responsible for maintaining all off-site facilities.

b. The customer shall be responsible for maintaining all on-site facilities (downstream of the user's service meter).

Facilities Design and Construction

a. Plans and specifications for all dual water distribution facilities shall be submitted to and approved by the appropriate regulatory agencies prior to construction.

b. Materials, equipment, and design and construction methods must be in compliance with local governing codes, rules, and regulations.

c. Prior to commencement of service to any secondary water system, record drawings shall be provided and approved and the installed system shall be tested under active conditions to ensure that the operation is acceptable.

d. A contingency plan shall be developed indicating procedures to be followed in the event that the nonpotable water does not meet applicable standards and a means for assuring a safe and continuous supply at all times.

Off-site Facilities

Operation and surveillance of all the off-site secondary water system facilities, including but not limited to pipelines, reservoirs, pumping stations, fire hydrants, manholes, valves, connections, supply interties, treatment facilities, and other appurtenances and property up to and including the meter, shall be under the management and control of the purveyor. No other persons except authorized employees of the purveyor shall have any right to enter upon, inspect, operate, adjust, alter, or move any portion of the foregoing or any of the purveyor's property. In the event that such should occur, all charges and penalties shall be applicable and collected.

On-site Facilities

a. The operation and surveillance of on-site water distribution facilities are the responsibility of the customer.

b. The operation and surveillance of all on-site secondary water system facilities shall be under the management of the water supervisor designated by the customer and approved by the purveyor and regulatory agencies. The water supervisor shall review all plans for on-site uses of nonpotable water and shall monitor these on a continuing basis. Authorized representatives of the purveyor shall monitor and inspect the entire water sytem, including on-site and off-site facilities, and for these purposes shall have the right to enter upon the customer's premises during reasonable hours.

c. In the event of contamination or pollution of the drinking water system because of a cross connection on the premises, the local health officer and water purveyor shall be advised promptly by the person responsible for the water system so that appropriate measures can be taken to overcome the contamination.

d. The customer shall have the following responsibilities in relation to operation of on-site facilities:

1. To ensure that all operations personnel are trained and familiar with the use of the secondary water.

2. To furnish their operations personnel with maintenance instructions and record drawings to ensure proper operation of the facilities.

3. To notify the purveyor of any and all updates or proposed changes, modifications, or additions to the on-site facilities. Any changes must be approved by the purveyor and regulatory authorities. Such changes must be submitted for plan check and approval prior to construction.

4. To ensure that the secondary water facilities remain in compliance with all applicable regulations and standards.

5. To operate and control the system in order to ensure a reasonable degree of public health protection. The customer shall be responsible for any and all subsequent uses of the water.

6. To comply with any and all applicable federal, state, and local statutes, ordinances, regulations, contracts, and all requirements prescribed by the purveyor. In the event of violation, all charges and penalties shall be applied and collected.

Monitoring and Inspection

a. The purveyor or authorized representatives of the purveyor shall monitor and inspect the entire dual water system, including both on-site and off-site facilities.

1. Inspections shall include all parts of the distribution system during construction to ensure compliance with required construction methods, materials, and placement.

b. The purveyor shall conduct monitoring programs, maintain records as necessary, and provide reports as requested by regulatory agencies. The purveyor or authorized representatives of the purveyor, in carrying out these functions, shall have the right to enter upon the customer's premises during reasonable hours for the purposes of inspecting on-site dual distribution facilities and areas of secondary water use to ensure compliance with all appropriate standards and regulations.

1. All violations and deficiencies shall be reported immediately to the state and local health departments and state or regional water quality control agency.

c. State or local regulatory agencies shall develop and institute independent inspection and monitoring programs as they deem necessary.

Conclusion

Dual distribution systems are not widely used at this time. However, such systems should be a commonly considered alternative in the near future, and system designs should be generally available.

It is imperative to find an alternative to distributing only potable water for any and all purposes. Dual distribution appears to be a feasible alternative that should be considered by design engineers and community water system managers.

As this manual has shown, a combination of increasing water treatment costs and decreasing sources of quality raw water have underscored the necessity of considering dual distribution systems to meet future demand. Although dual distribution is not a new concept, it has been utilized only minimally. Until or unless an all-around better alternative is proposed, the dual distribution network should receive serious consideration from water supply and treatment professionals, federal, state, and local governments, and consumers.

Dual distribution is beginning to find its proper place in engineering planning. Through application of this concept quality sources can be conserved and economies achieved. Where appropriate, dual distribution systems can optimize financial and water resources.

INDEX